THE MOST BEAUTIFUL
MATHEMATICAL FORMULAS

Lionel Salem

Frédéric Testard

Coralie Salem

Translated by
James D. Wuest

John Wiley & Sons, Inc.
New York • Chichester • Weinheim • Brisbane • Singapore • Toronto

Copyright © 1990 by InterEditions
Copyright © 1992 by John Wiley & Sons, Inc.
Published by John Wiley & Sons, Inc.

This publication is designed to provide accurate and authoritative information in regard to the subject matter covered. It is sold with the understanding that the publisher is not engaged in rendering legal, accounting, or other professional services. If legal advice or other expert assistance is required, the services of a competent professional person should be sought.

Originally published as *Les Plus Belles Formules Mathématiques* with the following restrictions of use in France:

Library of Congress Cataloging-in-Publication Data

Salem, Lionel
    The most beautiful mathematical formulas / Lionel Salem, Frédéric Testard, Coralie Salem ; translated by James D. Wuest.
        p.   cm
    Includes index.
    ISBN 0-471-55276-3 (cloth)
    ISBN 0-471-17662-1 (paper)
        1. Mathematics—Formulae.   I. Testard, Frédéric, 1961–
    II. Salem, Coralie.   III. Titie.
    QA41.S25   1992
    510—dc20   91-31959

Printed in the United States of America

10 9 8 7

To Bethsabée, Sylvie,
and everyone, young and old,
who has a passion for mathematics

Contents

The Most Beautiful Mathematical Formulas

Contents

Introduction

The goal of this book is to reveal the beauty of mathematical formulas. This beauty springs from the plasticity of mathematical symbols, the simplicity of mathematical statements, and the esthetic appeal of their implications. Like all sciences, mathematics has its own special harmony. Our goal is to explore this harmony.

We try to explain this beauty in mathematically simple terms. The origin of each formula is described and its validity is demonstrated. Some concepts, such as the Pythagorean theorem and the area of a circle, have been chosen for their universality. Others, such as Fermat's last theorem and Goldbach's conjecture, are justified by their historical importance. Still others, including logarithms and the formulas of trigonometry, have been chosen because they are used every day by young students of mathematics.

To further illustrate the beauty of mathematics, we have created a series of drawings that evoke the joy of discovery and symbolize the profound significance of mathematical formulas. We have tried to make the text and drawings amusing without doing injustice to mathematical truths, since we firmly believe that mathematical activity involves a sense of play.

This playfulness has led us to make up stories involving real and imaginary characters. Among the real individuals are the mathematicians d'Alembert, Archimedes, Cardano, Euler, Fermat, Fibonacci, Gauss, Goldbach, Lagrange, Leibniz, Napier, Newton, Pascal, and Pythagoras, as well as the Greek philosopher Zeno of Elea, the French philosopher Diderot, Catherine the Great of Russia, and a U.S. president, James Garfield. Brief biographical sketches of this diverse group of individuals can be found on pages 133–138. The imaginary characters include Cosine, a somewhat distracted scholar borrowed from a book by Christophe, a French illustrator and novelist. We have given Cosine a more modern look and some abnormal physical traits that he did not have in Christophe's book. Other fictional characters include Cosine's colleague Professor Sine, as well as a clever Dutch bicycle racer and other unnamed individuals with mathematical talent, including country folk, gardeners, a poster lover who hangs signs, and a mischievous little girl who cuts up globes. Certain events and situations, including the race for the Golden Gouda and the discovery of the Egyptian artifacts, are complete fabrications. The reader can turn to the annex at the end of the book for additional information that allows the real events and characters to be placed in their proper historical context, as well as for further information about the situations that we have made up.

The various sections of this book present a broad range of mathematical concepts that vary widely in difficulty. The chapters are arranged in an essentially spontaneous order, following what appears to us to be a natural progression of subjects. In part, this order reflects the way the authors think, and it is certainly not the only possible choice. As a result, subjects do not necessarily appear in order of increasing difficulty. For example, chapters in the middle of the book that treat logarithms and exponentials are less easily accessible than later chapters on spatial objects and numbers. Each section is largely independent of the others, so the reader is free to browse through the book at will. Because we want the book to be accessible, we have never let the rigor of proofs and vocabulary take precedence over the basic understanding of concepts. In fact, an explanation that is perfectly rigorous and exact is invariably more long and difficult, and it often obscures more than it clarifies.

The book is part of the intellectual legacy of Raphaël Salem, whose ideas and style have influenced many contemporary mathematicians.

We give our warm thanks to Professors Jean-Pierre Kahane and Henri Cartan. Professor Kahane gracefully prodded us to write the book when it was still just an idea, and we are very

Introduction

grateful to him. Professor Cartan agreed to read our drafts and contributed significantly to the rigor and clarity of the text by correcting errors and suggesting simplifications. We would also like to thank Jean-Pierre Sicre whose critical comments, based on extensive teaching experience, allowed us to make the most difficult parts of the book more accessible. We are deeply grateful to Geoffrey Staines for having faith in us. Finally, the book would never have been produced without the crucial assistance of Martine Wiznitzer and Pascaline Jay.

In conclusion, we hope that the beauty revealed by this brief stroll through the world of mathematics will inspire the reader to go further and continue to explore new realms.

Lionel Salem
Frédéric Testard
Coralie Salem

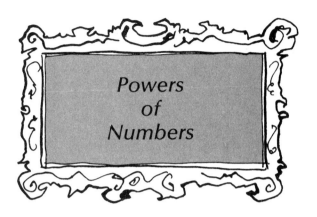

Powers
of
Numbers

1

Whole Powers of Numbers

If we take a number a, we call a multiplied by itself the square of a, and we write it a^2:

$$a^2 = a \times a$$

If $a = 3$, then $a^2 = 3 \times 3 = 9$. The word *square* comes from the fact that if we calculate the area of a square with side a, the result is a^2. For example, if $a = 3$, then the square can be subdivided into $3 \times 3 = 9$ little squares of side 1, whose area is 1 by definition. Thus the serving tray in the figure is ready to hold 9 pieces of cake.

In the same way, a multiplied by itself twice is the cube of a, written a^3:

$$a^3 = a \times a \times a$$

For example, $3^3 = 3 \times 3 \times 3 = 27$. This time, the word *cube* takes account of the fact that a^3 is the volume of a cube of side a, which can be subdivided into a^3 little cubes of side 1 and volume 1. Thus the culinary extravaganza in the figure consists of 27 pieces of cake.

In general, we define

$$a^4 = a \times a \times a \times a$$
$$(3^4 = 3 \times 3 \times 3 \times 3 = 81)$$
$$a^5 = a \times a \times a \times a \times a$$
$$(3^5 = 3 \times 3 \times 3 \times 3 \times 3 = 243)$$

and so on. In these cases there is no longer a simple geometric interpretation.

Finally, we can also define *negative powers* of numbers: The number a^{-1} is equal, by definition, to the reciprocal $1/a$ of a. For example, $2^{-1} = 1/2 = 0.5$; $2^{-2} = 1/2^2 = 1/4 = 0.25$; $2^{-3} = 1/2^3 = 1/8 = 0.125$; and so on.

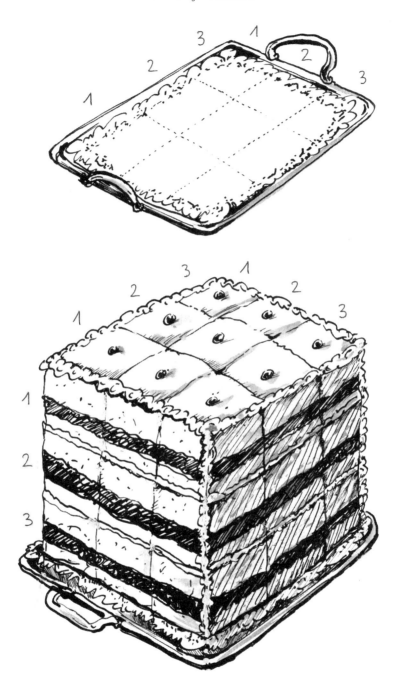

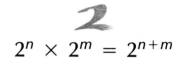

$$2^n \times 2^m = 2^{n+m}$$

We have just seen how to "raise a number to a given power." To raise the number a to the square is to calculate $a^2 = a \times a$; to raise to the cube is to calculate $a^3 = a \times a \times a$; and so on.

For example, the number

$$2^5 = 2 \times 2 \times 2 \times 2 \times 2$$

can be calculated by multiplying together five numbers all equal to 2. If we group the five numbers into a set of three and a set of two, we see that

$$2^5 = (2 \times 2 \times 2) \times (2 \times 2)$$
$$2^5 = 2^3 \times 2^2$$

Since the number of ancestors doubles with each generation (2 parents, 4 grandparents, and so on), we know that the number of ancestors in the fifth generation will be 2^5, which can be calculated by writing

$$2^5 = 2^3 \times 2^2$$
$$= 8 \times 4$$
$$= 32$$

Since each of the four grandparents has eight great-grandparents, the young rocker in the figure has 32 ancestors in the fifth generation, at the time of the first waltzes.

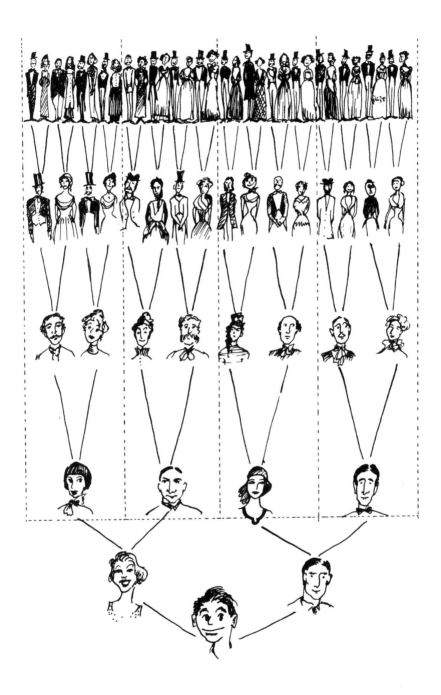

Triangles,
Rectangles,
Squares,
and Circles

3

The Area of a Rectangle Is Equal to the Product of its Sides

As the result of a bizarre mutation, certain four-leaved clovers have turned into giant plants, hastily baptized *Trifolium giganteum* by astonished botanists. These remarkable *Trifolia gigantea* have the strange property of growing only when they occupy a space of 1 meter by 1 meter, or 1 square meter.

A farmer who wants to raise these curious plants happens to have a rectangular field measuring 5 meters by 7 meters. He sees that he can sow five rows separated by one meter, with seven plants in each row. He therefore cultivates

$$5 \times 7 = 35$$

plants, without wasting the least bit of his field. Its *area*, a term derived from the Latin word for a piece of level ground, is therefore equal to 35 square meters, the product of its length and width.

$$A_{\text{Rectangle}} = a \times b$$

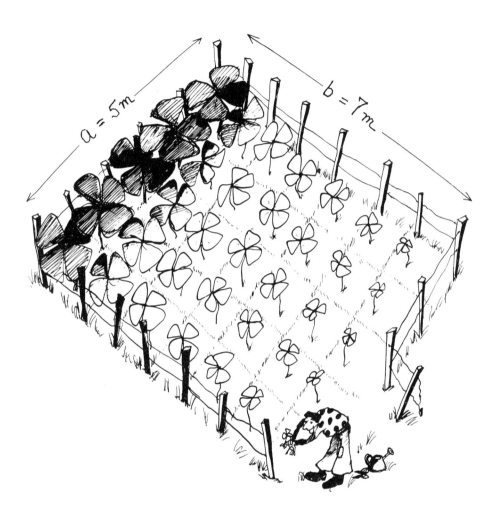

The Area of a Triangle Is Equal to One-half the Product of Its Height and Base

Raising *Trifolium giganteum* proves to be a major commercial success, and our farmer would very much like to grow more. The only field he has left is a triangular plot. Unfortunately, by another strange genetic twist of fate, the mutant plant will grow only in rectangular fields.

"No problem!" shouts the farmer. "A triangle is only half of a rectangle." He then goes to see his neighbor and suggests a deal.

"In my triangular field," he explains, "I can draw a line h perpendicular to the base b that passes through the opposite vertex. This cuts the field into two right triangles. If we join forces, we can turn the field into a rectangle for growing *Trifolium*, since we only need to double each of the triangles."

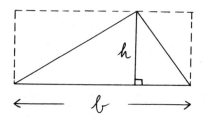

"Since the area of the rectangle is $b \times h$, the area of my triangle must be 1/2 ($b \times h$). So let's join our fields together and share the harvest equally."

$$A_{\text{Triangle}} = \frac{1}{2} (b \times h)$$

The Sum of the Angles of a Triangle Equals 180°

Excited by the idea of raising his first crop of *Trifolium*, the neighbor anxiously paces around the perimeter of his triangular field while awaiting the return of his partner, who has gone to get seeds. The neighbor baptizes the triangle ABC and notices that if the sides AC and AB form an angle α, then each time he reaches A he must turn through an angle of $180° - \alpha$. This is because the sum of the *external* and *internal* angles at A is equal to 180°. For the same reason, he turns $180° - \beta$ at B and $180° - \gamma$ at C.

After the three turns he finds himself pointing in the same direction, but he has made a complete circuit of 360°. He therefore realizes that

$$(180° - \alpha) + (180° - \beta) + (180° - \gamma) = 360°$$

or that

$$\boxed{\alpha + \beta + \gamma = 180°}$$

For further amusement, the reader can derive the equations that the farmer would have discovered if his field had had 4, 5, 6, or more sides.

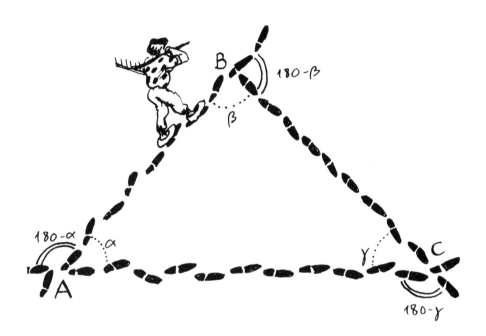

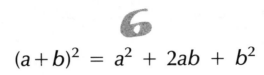

$$(a+b)^2 = a^2 + 2ab + b^2$$

One fine day, a whiz kid notices that the splendid poster of Van Gogh's "Sunflowers" hanging in the hallway of his parents' house has four creases. Intrigued, he looks more closely and sees that the creases divide the square poster into two square subsections, with sides a and b, and two rectangular subsections, with length a and width b. He then decides to entertain himself by calculating the area of the whole poster in two different ways.

He immediately sees that it must be $(a+b)^2$. But it is also $a^2 + b^2 + ab + ab$, since the area of the poster is the sum of the areas of the four subsections. Just to make sure, he decides to cut it up along the creases. After a few quick snips of his scissors, he confidently concludes that

$$\boxed{(a+b)^2 = a^2 + 2ab + b^2}$$

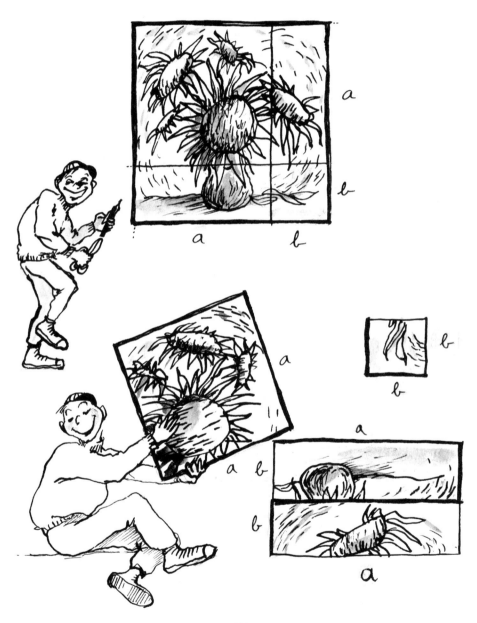

$$(a + b)(a - b) = a^2 - b^2$$

Our whiz kid has grown up in the streets and now pastes ads on billboards for a living. Having never lost his passion for dabbling in mathematics, he finds many opportunities in his new line of work to put his interest in formulas to good use. From his early experience cutting up posters, he has now moved on to explore new problems. He starts to hang a square sign with sides a, which has a small square subsection with sides b. He then decides to cut out the smaller square. This leaves a truncated L-shaped sign, which clearly must have an area of

$$a^2 - b^2$$

On a hunch, he moves the upper part of the L, which has sides of b and $a - b$, and places it on the bottom right end of the base of the L. He notices that this produces a large rectangle with length $a + b$, height $a - b$, and area

$$(a + b)(a - b)$$

This means that

$$(a + b)(a - b) = a^2 - b^2$$

Then he takes off for the day, telling himself that he has discovered a truly remarkable equality.

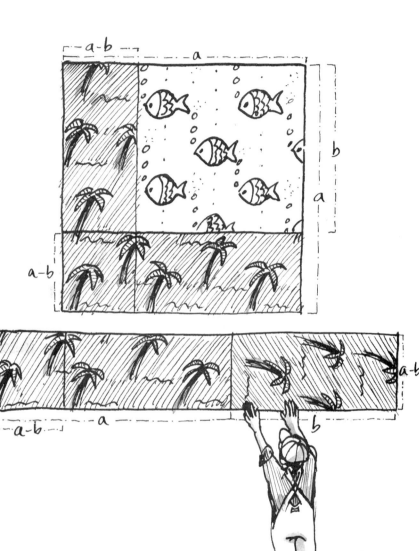

The Pythagorean Theorem:
$a^2 + b^2 = c^2$

To prove that in a right triangle the square of the hypotenuse is equal to the sum of the squares of the sides,

$$a^2 + b^2 = c^2$$

the U.S. president James Garfield decided to construct the trapezoid shown in the figure, which consists of two right triangles with sides a, b, and c, and a right isosceles triangle with two equal sides of length c. It is possible to verify that the angle AOB, which is just $180° - \alpha - \beta$, must be equal to $90°$, since the farmer in Chapter 5 has shown that the sum of the angles $\alpha + \beta + 90°$ in each right triangle is $180°$.

President Garfield then calculated the area of the trapezoid in two different ways. He was aware that the area is simply the product of its base, $a + b$, times one-half the sum of its sides, $1/2$ $(a+b)$. We have not proved this, but the reader can easily show that it is true by building a rectangle from two trapezoids.

The area can also be calculated by adding the areas of each of the three triangles, $ab/2$, $ab/2$, and $c^2/2$. As a result,

$$(a+b) \times \frac{(a+b)}{2} = ab + \frac{c^2}{2}$$

By using what the billboard specialist learned in Chapter 6 while he was an apprentice, President Garfield could expand the left side of the equation and rediscover the equality found almost 2500 years earlier by Pythagoras.

$$\boxed{a^2 + b^2 = c^2}$$

President Garfield proving the Pythagorean theorem with the help of a young poster lover.

9
The Circumference of a Circle
Equals $2\pi R$

A cyclist notices that it requires more effort to make one full turn of his wheels when their diameter is 70 centimeters than when it is 50 centimeters, and he decides to get to the bottom of the mystery. To measure the length of a full revolution, he takes each of his tires, puts a little dab of paint on the treads, rotates the tires, and then measures the distance between the spots of paint. For the 50-centimeter tire he measures a distance of 1.57 meters, and for the 70-centimeter tire he measures 2.19 meters. Repeating this experiment with tires of different diameters, he finds that their circumference L is always proportional to their diameter D and that the coefficient of proportionality is approximately 3.14.

$$L \approx 3.14 \times D$$

Somewhat later, he discovers this formula in a book on geometry. There the coefficient of proportionality is called π, which he learns comes from the first letter of the Greek word for *perimeter*. The diameter can be replaced in the formula by $2R$, twice the radius.

$$\boxed{L = 2\pi R}$$

In the book, our cyclist also finds the definition of another unit for measuring angles, called a *radian*. A radian, which is approximately 57° 18′, is the angle that cuts a length of 1 on the circumference of a circle of radius 1.

Since the length of one-fourth of a circle of radius 1 is $\pi/2$, $\pi/2$ is the value of a right angle in radians. Similarly, an angle of 180° corresponds to π radians, since half the circumference of a circle of radius 1 is π. More generally, in a circle of radius 1 the arc cut by an angle of α radians has a length of α. In particular, an angle of 360°, which corresponds to a full circle, measures 2π radians since the circumference of a circle of radius 1 is 2π.

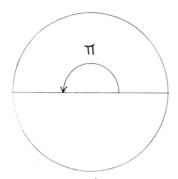

Angle of π radians
(180°)

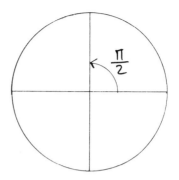

Angle of $\frac{\pi}{2}$ radians
(90°)

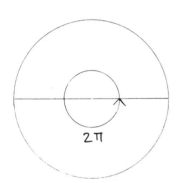

Angle of 2π radians
(360°)

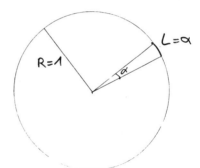

In a circle of radius 1, the arc cut by an angle of α radians has a length of a.

10

The Area of a Circle Equals πR^2*

In about 250 B.C., a pastry chef in Syracuse baked a huge round cake and invited everyone in town to come share it. After it had been cut up, which took several hours, each guest received a small slice. A number of guests noticed that the slices resembled right triangles with height R and base l, as shown in the drawing at the top of the opposite page.

The area of each slice was therefore approximately $1/2\,(R \times l)$. But one clever guest, Archimedes, pointed out that if the areas of all the little slices were added together, the result would be the radius R times the sum of all the lengths l divided by 2, which would be equal to the area A of the whole cake.

"However," he added, "the sum of the lengths l is just the circumference of the cake, which is $2\pi R$." The guests claim that he then shouted "Eureka!" and wrote in the icing

$$A = \frac{1}{2}R \times 2\pi R$$

or

$$\boxed{A = \pi R^2}$$

---

*Mathematicians make a subtle distinction between a "circle," which has a circumference but no area, and a "disk," which corresponds to the area inside the circle. In keeping with popular usage, however, we have retained the word "circle."

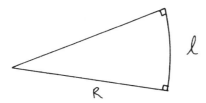

Angles

11

$$\cos^2 \alpha + \sin^2 \alpha = 1$$

Around 1900, the great-grandparents of the authors of this book got Christophe, the French writer and illustrator, to introduce them to one of his most elusive friends, the great scholar Cosine. After a long, blindfold trip in one of the first automobiles, our great-grandparents finally arrive at the home of the master. At his side is Professor Sine, introduced as his most faithful collaborator.

Encountering Professor Sine is the first of many surprises. Our great-grandparents quickly see that the scholar Cosine and Professor Sine are both dwarf contortionists, able to shrink from one meter to less than a centimeter and then grow back again upside-down. The two of them exchange conspiratorial winks and announce to the guests that they will proceed to give them a lesson in trigonometry.*

"You know, of course," they begin, "that in a right triangle the sine of an angle α is calculated by dividing the opposite side by the hypotenuse, and the cosine is obtained by dividing the adjacent side by the hypotenuse, as the drawing on the blackboard indicates. As you can see, however, this definition does not give the sine or cosine of obtuse angles."

"Fortunately," they add, "there is a way to define these sines and cosines." They lie down on two lines, one vertical and the other horizontal, defining two diameters of a circle of radius 1. Then they ask their visitors to choose an angle. Our great-grandparents pick 30°. The scholar Cosine adjusts his height to the cosine of 30°, about 0.866, and Professor Sine matches his

---

*The term *trigonometry* is from the Greek *trigônos* for triangle and *metron* for measure.

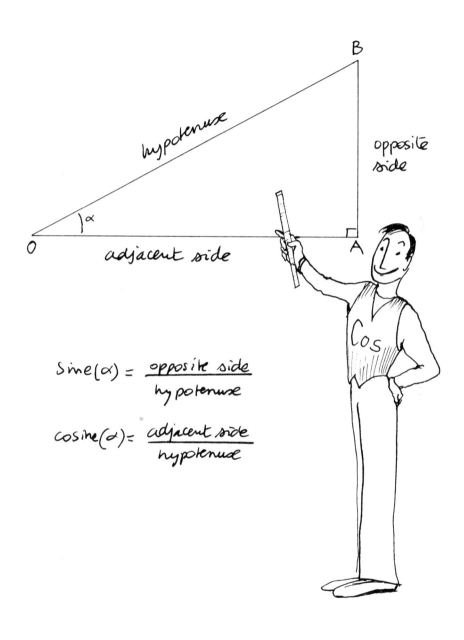

$$\text{Sine}(\alpha) = \frac{\text{opposite side}}{\text{hypotenuse}}$$

$$\text{cosine}(\alpha) = \frac{\text{adjacent side}}{\text{hypotenuse}}$$

height to the sine of 30°, which is 0.5. "Note carefully," says the scholar Cosine, "that in the right triangle OAB the hypotenuse is equal to 1 and the 'new' sine and cosine are identical to the normal values."

When our great-grandparents pick 90°, the head of Professor Sine touches the circle, and the scholar Cosine becomes so small that he can no longer be seen. This means that

$$\sin 90° = 1, \cos 90° = 0$$

On the other hand, when 180° is picked, it is Professor Sine who disappears while the scholar Cosine attains his maximum height. This time, however, he lies on the left side of the line, which means that the cosine is negative:

$$\sin 180° = 0, \cos 180° = -1$$

"You see, then," they conclude, "that in this way it becomes possible to calculate the sine and cosine of an obtuse angle."

A moment later, they abruptly ask their guests if they know why

$$(\cos \alpha)^2 + (\sin \alpha)^2 = 1$$

for all values of α. In keeping with mathematical tradition, this can also be written $\cos^2 \alpha + \sin^2 \alpha = 1$. After a few minutes, their guests give up. Sine and Cosine then explain that each of them represents the side of a right triangle OAB whose hypotenuse is equal to 1, the radius of the circle. It is therefore obvious from the Pythagorean theorem that the square of one side plus the square of the other equals the square of the hypotenuse, which is 1.

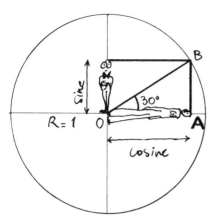

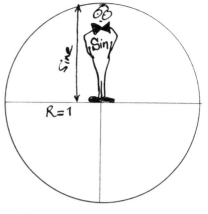

1) Angle of 30°
$$\text{sine} = \frac{1}{2}$$
Cosine ≥ 0.866

2) Angle of 90°
Sine = 1
Cosine = 0

3) Angle of 180°
Sine = 0
Cosine = 1

4) $Cos^2 \alpha + sin^2 \alpha = 1$

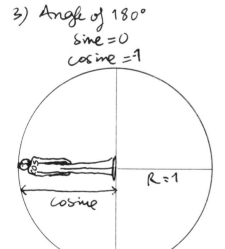

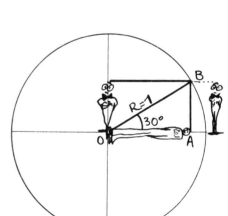

12

$$\sin \alpha \approx \alpha, \cos \alpha \approx 1 - \frac{\alpha^2}{2}$$

(α small)

"Since we do not own a calculator," they complain, "we need to find a way to figure out, at least approximately, the sine and cosine of a given angle. The length of an arc of a circle of radius 1 is equal to the angle α when the angle is expressed not in degrees but in radians. A radian, which is equal to about 57°, is another unit for measuring angles that we have already encountered in Chapter 9. If the arc is very small, then it will be practically identical to the straight line that subtends it, which is in turn very similar in length to the vertical line that completes a right triangle in the figure on the opposite page." This line has the same length as Professor Sine, so

$$\boxed{\sin \alpha \approx \alpha}$$

Our great-grandparents are already totally amazed, but the scholar Cosine presses on: "Now look at the little shaded right triangle at the right of Professor Sine. Its sides are exactly $\sin \alpha$ and $1 - \cos \alpha$ and its hypotenuse is approximately α."

By using the Pythagorean theorem and the formula provided by the poster lover in Chapter 6, we find that

$$\sin^2 \alpha + (1 - \cos \alpha)^2 \approx \alpha^2$$
$$\sin^2 \alpha + 1 + \cos^2 \alpha - 2 \cos \alpha \approx \alpha^2$$

But since $\sin^2 \alpha + \cos^2 \alpha = 1$, we have

$$2 - 2 \cos \alpha \approx \alpha^2$$

$$\cos \alpha \approx 1 - \frac{\alpha^2}{2}$$ *

"Be careful!" warns the scholar Cosine. "These are approximations valid only for small angles α."

---

*To be mathematically rigorous, this expression should be written $1 - \cos \alpha \approx \frac{\alpha^2}{2}$, and it should be understood to mean not an equivalence but a limiting value for $\cos \alpha$ as α approaches 0.

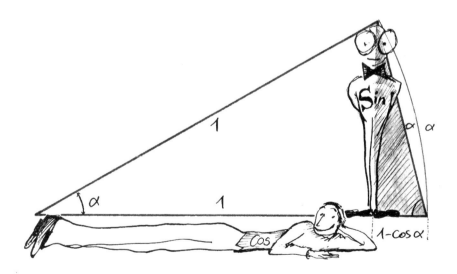

$$\frac{\sin \alpha}{a} = \frac{\sin \beta}{b} = \frac{\sin \gamma}{c}$$

13

"We are now going to show you something truly amazing," continue the scholar Cosine and Professor Sine. "You must not forget that in a right triangle

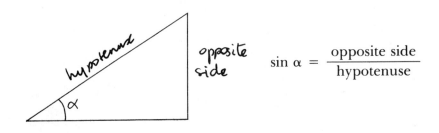

$$\sin \alpha = \frac{\text{opposite side}}{\text{hypotenuse}}$$

Let's take a triangle—perhaps this one here—and cut it vertically into two halves. This gives us two right triangles. We will call the angles at A and B α and β, the lengths of the opposite sides a and b, and the height h." Then

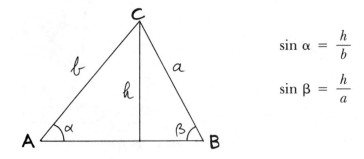

$$\sin \alpha = \frac{h}{b}$$

$$\sin \beta = \frac{h}{a}$$

So $h = b \sin \alpha = a \sin \beta$. From this equality,

$$\frac{\sin \alpha}{a} = \frac{\sin \beta}{b}$$

But since nothing prevents us from cutting the triangle between A and C, it must also be true that

$$\boxed{\frac{\sin \alpha}{a} = \frac{\sin \beta}{b} = \frac{\sin \gamma}{c}}$$

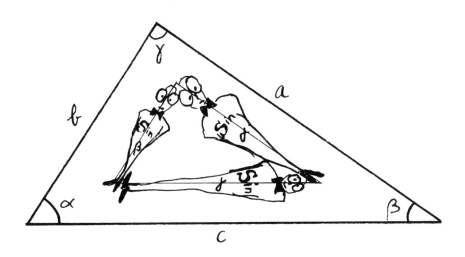

Rational and Irrational Numbers

Rational numbers are those obtained by dividing one whole number by another. For example, 2/3 (2 divided by 3) is a rational number. However, some numbers cannot be expressed as the quotient of two whole numbers. These numbers are called *irrational*. For example, the irrational numbers include $\sqrt{2}$, which is the length of the diagonal of a square of side 1; π, which is the circumference of a tire of diameter 1 (see Chapter 9); and the number e, which we will encounter in Chapter 23.

In the 18th century, during the traditional country fair in Amstel, a small town in the Netherlands, participants in the Golden Gouda bicycle race were asked not to go fast but to

construct bicycles with tires able to travel 22 meters in a whole number of revolutions. No one was ever able to claim the prize.

One year, Rik van Loy-Ertanet, a competitor a little more clever than the others, decided that with a few calculations and a bit of determination he would be able to win. He began by using a tire one meter in diameter, but he was disappointed to find that because its circumference was equal to π, a whole number of revolutions could never equal 22, since π is irrational. He was greatly frustrated to find that after seven full revolutions he was just 9 millimeters short of 22 meters.

The next year, he reviewed his calculations and constructed a tire whose circumference measured 22/7. He was then able to cover the distance of 22 meters in exactly 7 full revolutions. This is because 22/7, like all rational numbers, has multiples that are whole numbers. In this case, $7 \times (22/7) = 22$.

Although he kept his discovery strictly to himself, the secret eventually got out, and there were more and more co-winners of the prize in each subsequent year. Finally the organizers got tired of trying to settle ties, so they changed the rules of the competition: Henceforth, the winner would be the fastest. In this way the modern bicycle race was born.

$$\pi \approx \frac{355}{113}$$

We have just seen that there is no way to express π as the ratio of two whole numbers. However, there are a variety of approximations of this type, including $3.14 = 314/100 = 157/50$. Here is a clever method for finding another:

$$\pi \approx 3.1415926 = 3 + 0.1415926$$

We then can write

$$0.1415926 \approx \frac{1}{(7.062516)} = \frac{1}{(7 + 0.062516)}$$

Since

$$0.062516 \approx 1/15.99$$

and since we can round off 15.99 to 16, we then have

$$\pi \approx 3 + \cfrac{1}{7 + \cfrac{1}{16}}$$

$$= 3 + \cfrac{1}{\cfrac{113}{16}} = 3 + \frac{16}{113} = \frac{355}{113}$$

We find that

$$355/113 \approx 3.1415929$$

whereas

$$\pi \approx 3.1415926$$

This means that the difference between π and the approximation given by 355/113 is only about 3 ten-millionths.

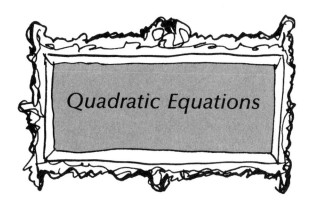

Quadratic Equations

16

The Roots of a Quadratic Equation

Around 500 B.C., a Greek farmer wanted to buy a square field and something to enclose it with. Each square meter of ground cost a drachma, as did each meter of fence. He had a total of 60 drachmas. He calculated that if the square measured N meters on a side, then its area would be N^2 square meters and its perimeter would be $4N$ meters, so that the total cost for land and fence would amount to $N^2 + 4N$ drachmas. The largest field he could acquire would therefore be one for which

$$N^2 + 4N = 60$$

To solve this equation, he noticed that $N^2 + 4N$ is the "start" of $(N + 2)^2$ since $(N + 2)^2 = N^2 + 4N + 4$. (In arriving at this insight, he may have been helped by contemporary poster lovers and billboard specialists. See Chapter 6.) The equation then became

$$(N + 2)^2 = N^2 + 4N + 4 = 64$$

so $N + 2 = 8$, or $N = 6$. The other solution, $N + 2 = -8$, made no sense to the farmer.

Later, other mathematicians found that numbers N for which

$$aN^2 + bN + c = 0$$

exist only when $b^2 - 4ac \geq 0$ and are given by the formulas

$$N_1 = \frac{1}{2a}\left(-b + \sqrt{b^2 - 4ac}\right)$$

$$N_2 = \frac{1}{2a}\left(-b - \sqrt{b^2 - 4ac}\right)$$

If $b^2 - 4ac = 0$, the two solutions are identical.

The numbers N_1 and N_2 are called the *roots* of the preceding quadratic equation. The reader can try to derive these formulas by dividing the equation by a and proceeding in the same way as the Greek farmer.

17

The Golden Ratio: $\dfrac{1 + \sqrt{5}}{2}$

Since the days of old, artists as well as geometricians have known that there is a special, aesthetically pleasing rectangle with width 1, length x, and the following property: When a square of side 1 is removed, the rectangle that remains has the same proportions as the original rectangle (see the drawing on the opposite page). Since the new rectangle has a width of $x - 1$ and a length of 1, the equivalence of the proportions means that

$$\frac{x - 1}{1} = \frac{1}{x}$$

The number x, which is called the golden ratio, must therefore satisfy the equation

$$x^2 - x - 1 = 0$$

The formulas of Chapter 16 can be used to show that the positive root of this equation is

$$x = \frac{1 + \sqrt{5}}{2} \approx 1.618$$

As we did for the number π in Chapter 15, we will now try to find a rational number that approximates the irrational number x. To do this, we note that the first equation allows us to write that

$$x = 1 + \frac{1}{x}$$

On the right side of this equation, we can replace x by the equivalent expression $1 + \dfrac{1}{x}$. This gives the new equation

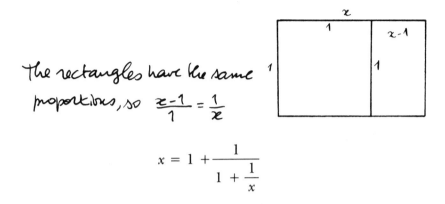

The rectangles have the same proportions, so $\dfrac{x-1}{1} = \dfrac{1}{x}$

$$x = 1 + \cfrac{1}{1 + \cfrac{1}{x}}$$

By carrying out further substitutions of the same kind and by omitting the last $1/x$, we arrive at the following series of numbers:

$$1,\ 1 + \frac{1}{1},\ 1 + \cfrac{1}{1 + \cfrac{1}{1}},\ 1 + \cfrac{1}{1 + \cfrac{1}{1 + \cfrac{1}{1}}},\dots$$

In practice, it is tedious to continue to calculate these numbers indefinitely, but it is nevertheless possible to see that they get closer and closer to the exact value of

$$\frac{1 + \sqrt{5}}{2}.$$

To represent this phenomenon of convergence, we can write

$$\frac{1 + \sqrt{5}}{2} = 1 + \cfrac{1}{1 + \cfrac{1}{1 + \cfrac{1}{1 + \cfrac{1}{1 + \dots}}}}$$

" the Annunciation," by Leonardo da Vinci.

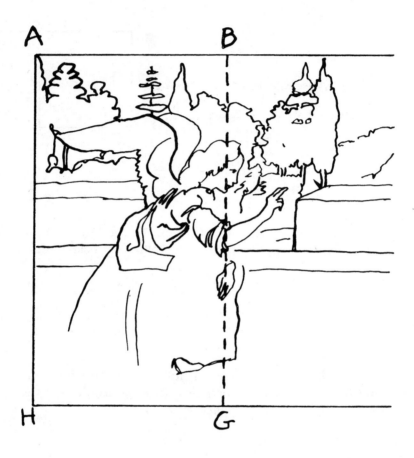

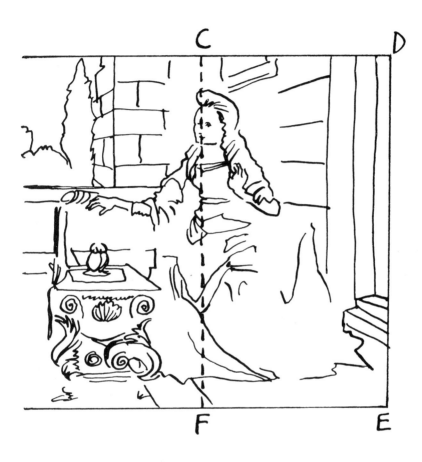

the two figures divide the painting into four
golden rectangles : ACFH,
                     BDEG,
                     ABGH, and
                     CDEF.

18

Imaginary Numbers: $i = \sqrt{-1}$

We learned how to solve $x^2 + 4x = 60$ and other quadratic equations in Chapter 16. The formulas given there allow us to show that the equation

$$x^2 - 3x + 2 = 0$$

has the roots $x_1 = 1$ and $x_2 = 2$. However, some quadratic equations do not have normal roots. The most simple example is the equation

$$x^2 + 1 = 0$$

for which the formulas of Chapter 16 give

$$x_1 = \frac{\sqrt{-4}}{2}, \qquad x_2 = -\frac{\sqrt{-4}}{2}$$

But the number $\sqrt{-4}$ does not exist, since no real number can have the negative number -4 as its square. In a sense, the existence of quadratic equations of this type forced mathematicians to invent *imaginary* numbers whose squares are negative. This was done in the 16th century by the Italian mathematician Girolamo Cardano, who developed formulas that could be used to solve cubic equations such as

$$x^3 - 13x + 12 = 0$$

Even when simple roots exist (in the preceding case, for example, they are 1, 3, and -4), Cardano was surprised to find that his formulas forced him to use the square roots of negative numbers. He therefore decided to accept the existence of such numbers. The simplest, called *i*, is the square root of -1:

$$i = \sqrt{-1}$$

This means that the roots of $x^2 + 1 = 0$ are *i* and $-i$.

By combining real numbers like 5 with imaginary numbers like $2i$, mathematicians have created a series of new numbers, such as $5 + 2i$, which are called *complex* numbers. In this case, 5 is called the *real* part of $5 + 2i$, and 2 is the *imaginary* part.

Imaginary number

Real number

*Logarithms
and
Exponentials*

19
The Discovery of Logarithms

Once upon a time, in 1614, Lord Napier had a Scottish gardener. Lord Napier was fascinated by geometry and extremely eccentric, and he made his employee plant gardens of every imaginable shape. In addition, the gardener had to systematically calculate the area of each one, since Napier would never give him a single seed more than was absolutely necessary. His first job was to plant a rectangular garden of width 1 and length x, and Napier added the restriction that nothing should be planted in the first meter. The area A that actually needed to be seeded was therefore

$$A = 1 \times (x - 1) = x - 1$$

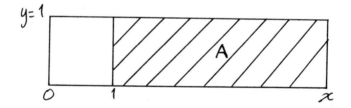

Then Napier asked him to plant a triangular garden bordered by the straight line $y = x$, with the same odd restriction. Remembering the formula for the area of a triangle, the gardener calculated that the planted area should be

$$A = \frac{x^2}{2} - \frac{1}{2}$$

which is just the area of the large triangle minus the area of the small one with side 1.

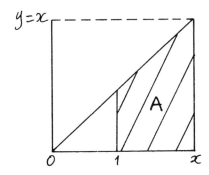

Ever more demanding, Lord Napier then insisted that the gardener plant a parabolic garden, bordered this time by the curve $y = x^2$. He opened his mouth to continue to speak, but the gardener interrupted him. "I know," he said. "Nothing planted in the first meter."

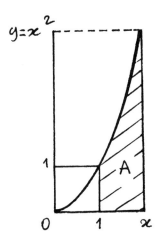

This time the gardener had to consult a geometry book to find that the planted area A should be

$$A = \frac{x^3}{3} - \frac{1}{3}$$

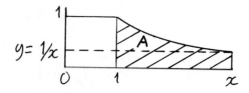

Getting caught up in the game, he then came up with the idea of planting a hyperbolic garden, which becomes narrower after the first meter instead of broader. This time, however, all his efforts to find a formula in a book were in vain. In honor of his employer, who had awakened his interest in geometry, the gardener therefore decided to call the area a Napierian surface. Later, mathematicians would rename it a *Napierian logarithm.* The logarithm of x, which is written log x or ln x, therefore corresponds to the cross-hatched area A shown in the drawing above:

$$A = \log x$$

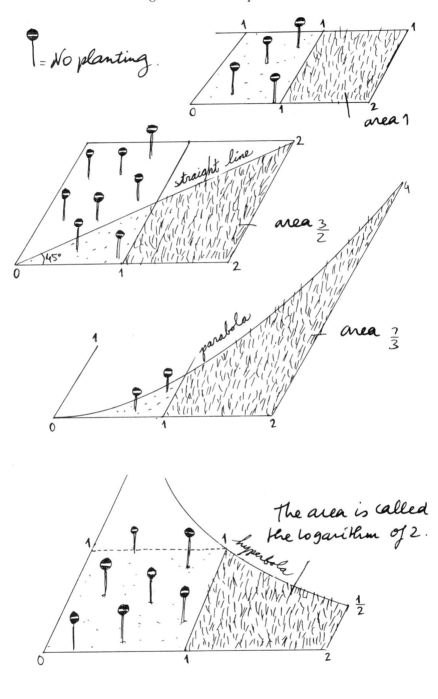

= No planting.

area 1

straight line

45°

area $\frac{3}{2}$

parabola

area $\frac{7}{3}$

hyperbola

The area is called the logarithm of 2.

$\frac{1}{2}$

20

The Wonderful Law of Logarithms:
log (*ab*) = log *a* + log *b*

It was one thing to give logarithms a name, but something else altogether to calculate them. The gardener quickly saw that when $x = 1$ the garden was reduced to a simple border and therefore had no area whatsoever:

$$\log 1 = 0$$

To calculate other values, he cut the area to be measured into a large number of roughly rectangular subsections. For example, the drawing on the opposite page shows how log 4 can be divided into three parts. By using this laborious method, which requires dividing the area into a larger and larger number of smaller and smaller subsections, and by following the procedure that we will see later in Chapter 22, he found that

$$\log 2 \approx 0.693$$
$$\log 3 \approx 1.098$$
$$\log 4 \approx 1.386$$
$$\log 5 \approx 1.609$$
$$\log 6 \approx 1.791$$

Suddenly the gardener gave a shout of triumph. He had just noticed that

$$\log 4 = \log 2 + \log 2$$
$$\log 6 = \log 2 + \log 3$$

Boldly, he deduced that for all numbers a and b

$$\boxed{\log (ab) = \log a + \log b}$$

Firm proof was provided somewhat later, thanks to the efforts of Isaac Newton, another Briton who drew inspiration from gardens, particularly ones with fruit trees.

21

$$1 + \frac{1}{2} + \ldots + \frac{1}{n} - \log n$$

Converges to 0.577...

Discovery of the magic formula

$$\log (ab) = \log a + \log b$$

made it easy for the gardener to calculate new values without needing to make further measurements. For example, to calculate log 18 he could use the fact that

$$18 = 2 \times 9$$

to write that

$$\log 18 = \log 2 + \log 9$$

But since $9 = 3 \times 3$,

$$\log 9 = \log 3 + \log 3$$

so finally

$$\log 18 = \log 2 + 2 \log 3 \approx 2.890$$

This method can be used to calculate even larger logarithmic values, and for entertainment the reader can show that log 100 or log 1000 can be expressed in terms of log 2 and log 5. However, the gardener was not convinced that the procedure could be used to calculate the values of all logarithms. By making drawings similar to those on the opposite page, he noticed that the special shape of the hyperbolic garden meant that the area

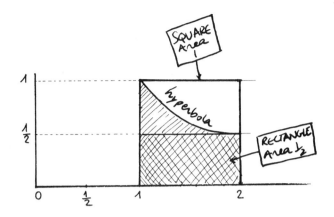

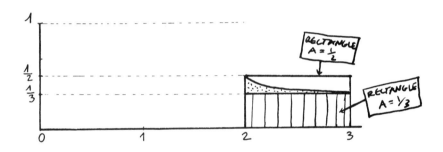

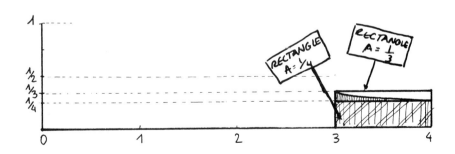

between the first and second meters was somewhere between 1/2 and 1, the area between the second and third meters was between 1/3 and 1/2, the area between the third and fourth meters was between 1/4 and 1/3, and so on.

He therefore decided to approximate the value of logarithms by computing the sum of the numbers 1, 1/2, 1/3, 1/4, and so on. For example, the approximate value of log 6 is

$$1 + \frac{1}{2} + \frac{1}{3} + \frac{1}{4} + \frac{1}{5} + \frac{1}{6} = 2.45$$

which is in error by

$$E \approx 2.45 - 1.791 = 0.659$$

In general, log n can be approximated by $1 + 1/2 + 1/3 + \ldots + 1/n$, which is just the sum of the reciprocals of all whole numbers from 1 to n. As n becomes larger and larger, it has been shown that this sum differs from the true value of log n by an amount C that is approximately equal to

$$\boxed{C \approx 0.577}$$

The number C is called Euler's constant in honor of a great mathematician. The formula is valuable because as n gets larger and larger, log n also gets larger and larger, and $1 + 1/2 + 1/3 + \ldots + 1/n - C$ provides an increasingly close estimate of the exact value of log n.

log (1 + x) ≈ x (x small)

It occurred to the gardener that it was silly to calculate big numbers like log 50,000 when he did not even know the value of smaller numbers like log (1.1). After resigning himself to making further measurements, he then had a flash of insight.

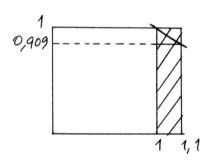

Between 1 and 1.1, the "height" of the garden ranges from 1 to 0.909, which is just the value of the hyperbolic curve $y = 1/x$ when $x = 1.1$ (see Chapter 19). The area under the curve is therefore somewhere between that of the vertical rectangle with sides 0.1 and 0.909 and that of the larger cross-hatched rectangle with sides 0.1 and 1. These two areas are 0.0909 and 0.1, respectively, so they differ by only 0.01. The gardener realized that he would be making only a small error if he estimated the area under the curve by taking the area of the larger rectangle; so the approximate value of log (1.1) is 0.1. In general, if x is smaller than 0.1, he could use x as the estimated value of log $(1 + x)$, since the area under the curve is approximately that of a rectangle of width x and height 1.

We have just seen in Chapter 21 that the formula log $(ab) =$ log a + log b allows us to calculate log 9, log 18, and other logarithms. By making use of the formula for estimating log (1

+ x) that we have just discovered, we can now calculate the logarithm of any positive number. Scientific calculators use a somewhat more precise version of this same procedure.

23
The Number e

Dazzled by all his discoveries, the gardener rushed to tell Lord Napier that he had created an extraordinary garden and that he would need a quantity of seeds to confirm his hypotheses. Lord Napier, as fussy as ever, then replied, "I will give you enough to plant 1 square meter, but no more!"

The gardener started back to work, somewhat irritated by Lord Napier's stinginess, but then he realized that he nevertheless had enough seeds to do something interesting. "Up to now," he said to himself, "I have chosen x and then calculated the area $\log x$. This time, though, let's do the opposite." He therefore decided to look for the value of x that would give an area $\log x$ equal to 1.

It is obviously a number between 2 and 3, he realized, since the area between 1 and 2 is smaller than 1 ($\log 2 \approx 0.693$ according to Chapter 20), whereas the area between 1 and 3 is larger than 1 ($\log 3 \approx 1.098$). After fiddling with the numbers, he finally found that the required value of x is approximately 2.718, and he christened this number e:

$$\boxed{e \approx 2.718}$$

Some suspect that e was chosen because it is the first letter of the word "England." In retaliation, mathematicians have shown that it is an irrational number (see Chapter 14).

The gardener begins to sow, starting with a full sack of seeds.

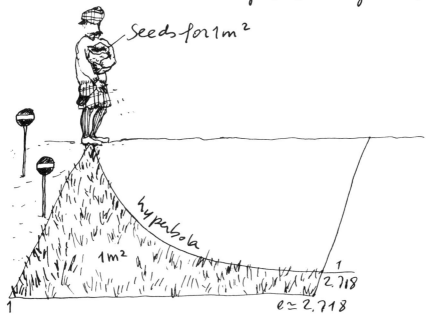

Seeds for 1 m²

hyperbola

1 m²

1

$e \simeq 2.718$

1
2.718

The sack is empty, and the garden is seeded.

24

The Number e Raised to a Real Power

Having found the number e for which $\log e = 1$, the gardener, always eager to go further, decided to look for the number a for which $\log a = 2$. He did the calculation and found that $a \approx (2.718)^2$. He then realized that the answer was obvious since if $\log a = 2 = \log e + \log e$, then according to the formula of Chapter 20 it must be true that $a = e \times e = e^2$.

If the gardener had picked a number x instead of 2, he would have needed to look for a number a for which $\log a = x$. It seemed reasonable to him to write this number as e^x, since he had just seen that when $x = 2$, $a = e^2$. In this way, the gardener invented "raising the number e to a real power." Such powers can even be irrational. Previously, we have only seen numbers raised to the square, the cube, or other whole powers as discussed in Chapter 1, but we can now write numbers like $e^{\sqrt{2}}$.

The gardener then decided to plant a new garden that would make use of the numbers e^x he had just discovered. This time Lord Napier removed his restriction on planting in the first meter, so the gardener seeded the entire surface. He was amazed to find that the area bounded by length x and height e^x is

$$A = e^x - 1$$

It is said that Lord Napier immediately offered him a well-deserved retirement.

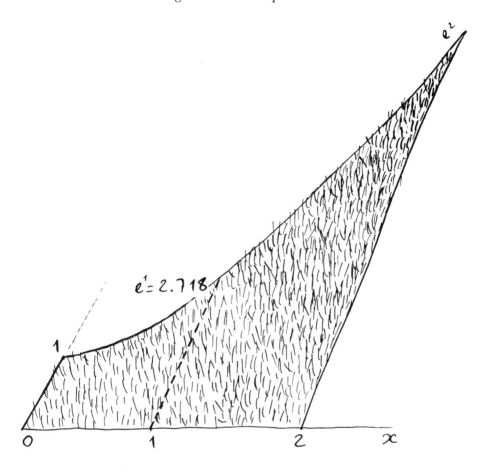

e^2

$e^1 = 2.718$

1

O   1   2 x

Area ⫿⫿ $= e^2 - 1$!!

25

Derivatives and Integrals: Areas Viewed from Two Different Perspectives

About sixty years after these events, Isaac Newton was sitting under an apple tree and reading an account of the adventures of Napier's gardener when he had a stroke of genius.

What had the gardener actually done? He had chosen the shape of the garden, either rectangular, triangular, parabolic, or hyperbolic, and then he had calculated the area between 1 and x. But couldn't the opposite be done too?

This is what Newton then set out to do. "Let's imagine," he said, "that the area between the points 0 and x is $x^2/2$. In this case, the garden must be bounded by the straight line $y = x$." He therefore decided to call x the "derivative" of $x^2/2$.

He only needed to read the account of the gardener's work to discover that the derivative of x is 1 (rectangular garden), that of $x^3/3$ is x^2 (parabolic garden), and that of $\log x$ is $1/x$. Finally, by using the last of the formulas, he found that the derivative of $e^x - 1$ is e^x.

Soon after, he realized that to double or triple the area of a garden, it was only necessary to double or triple its "height." As a result, if he wanted the derivative of x^3, he simply needed to calculate 3 times the derivative of $x^3/3$, or $3x^2$. Similarly, the derivative of x^2 equals $2x$. From there, he could quickly see that the derivative of x^4 is $4x^3$, that of x^5 is $5x^4$, and so on.

In the figures drawn at the bottom of the page, the upper line is a function of x. In general, this function defines the height of a surface that depends on x. We call the area of this surface the *integral* of the function, and the function itself is the *derivative* of this integral.

In fact, the word *derivative* did not actually come from Newton himself, who preferred the term *fluxion*, but is derived instead from the French mathematician Lagrange, who will appear in Chapter 40 in connection with a completely different story. Moreover, it was the great German mathematician and philosopher Leibniz who called $x^2/2$ the "integral" of x (elegantly written $\int x dx$), which gives the area of a surface bounded by the straight line $y = x$. By measuring areas, our gardener was calculating integrals without actually knowing it.

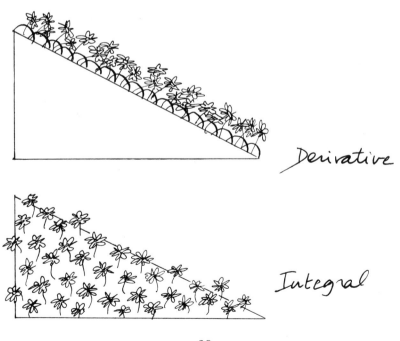

Derivative

Integral

26

The Number e Raised to An Imaginary Power: $e^{i\alpha} = \cos \alpha + i \sin \alpha$

The number e encountered in Chapter 23 has remarkable properties, much like the other special numbers i (Chapter 18) and π (Chapter 15). Even more extraordinary things happen when the effects of these numbers are combined.

We have already learned from Lord Napier's gardener how to raise the number e to a real power. Now, in a triumphant return to the scientific stage, our friends the scholar Cosine and Professor Sine will teach us about "imaginary powers" by defining

$$e^{i\alpha} = \cos \alpha + i \sin \alpha$$

for a given angle α.

"You probably feel," they immediately say, "that this hardly looks like the definition of a power." But notice that

$$e^{-i\alpha} = \cos (-\alpha) + i \sin (-\alpha)$$
$$= \cos \alpha - i \sin \alpha$$

so that

$$e^{i\alpha} \times e^{-i\alpha} = (\cos \alpha + i \sin \alpha)(\cos \alpha - i \sin \alpha)$$
$$= \cos^2 \alpha - i^2 \sin^2 \alpha$$
$$= \cos^2\alpha + \sin^2\alpha = 1$$

(We have not stated explicit rules for handling imaginary numbers for a very simple reason: They are identical to those for real numbers, although it is important not to forget that $i^2 = -1$). As a result,

$$e^{-i\alpha} = \frac{1}{e^{i\alpha}}$$

"As you can see," they conclude, "even though this power is totally imaginary, it behaves just like the real powers you have already seen at the end of Chapter 1."

Real part

Imaginary part

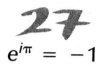

$$e^{i\pi} = -1$$

"As we have already stated," continue the scholar Cosine and Professor Sine, "amazing things happen when e, i, and π are combined." For example, remember that we have seen in Chapter 11 that

$$\cos \pi = -1, \qquad \sin \pi = 0$$

Under these conditions,

$$e^{i\pi} = \cos \pi + i \sin \pi = -1$$

Professor Sine informs us that this equality was discovered in the 18th century by the great Swiss mathematician Leonhard Euler. Professor Sine takes out an old print and explains to us that it portrays the famous conversation of Catherine the Great of Russia, Euler, and the philosopher Denis Diderot in which the empress asks if God exists. By long-winded and highly complex arguments, Diderot attempted to prove that God does not exist. In contrast, Euler simply stated, "Since $e^{i\pi} = -1$, God must exist!"*

"In honor of the great genius of Euler," continues Professor Sine, "his formula appears in the Museum of Discovery in Paris, in a room where π is written out to a huge number of decimals. If you would enjoy it, I'll take you there one of these days."

---

*The real story is even more hilarious. Euler made fun of Diderot by concocting a nonsensical formula:

"Sir, $\dfrac{a + b^n}{n} = x$, so God must exist! Answer that!"

Poor Diderot was too intimidated to reply (E. T. Bell, *Men of Mathematics*, Simon & Schuster, 1937).

28

$$\cos 2\alpha = \cos^2\alpha - \sin^2 \alpha;$$
$$\sin 2\alpha = 2 \sin \alpha \cos \alpha$$

Once again, we let the dynamic duo of Cosine and Sine guide us through the world of mathematics. Today, they introduce us to two of their students, named Cos II and Sin II, who are young contortionists able to change both their height and width.

"You surely remember," the scholar Cosine tells us, "that you encountered the number $e^{i\alpha}$ in Chapter 26. So let's see what $e^{2i\alpha}$ would be." First of all,

$$
\begin{aligned}
e^{2i\alpha} &= e^{i\alpha} \times e^{i\alpha} \\
&= (\cos \alpha + i \sin \alpha)^2 \\
&= \cos^2\alpha + 2\,i \sin \alpha \cos \alpha + i^2 \sin^2 \alpha
\end{aligned}
$$

(courtesy of our friend in Chapter 6)

$$= \cos^2\alpha - \sin^2\alpha + 2i \sin \alpha \cos \alpha$$

But we also know that

$$
\begin{aligned}
e^{2i\alpha} &= e^{i(2\alpha)} \\
&= \cos 2\alpha + i \sin 2\alpha
\end{aligned}
$$

When the real and imaginary parts are matched, two beautiful formulas emerge, which the students Cos II and Sin II now write down:

$$\cos 2\alpha = \cos^2\alpha - \sin^2\alpha$$
$$\sin 2\alpha = 2 \sin \alpha \cos \alpha$$

"To be completely honest," the scholar Cosine confides to us later, "it is possible to derive these two formulas without ever using imaginary numbers, but the proof is a bit more complicated. This shows how useful and interesting it can be to introduce abstract objects."

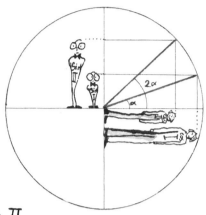

Area = sin II

$\alpha = 30°$
$\cos I \approx 0.866$
$\sin I = 0.5$
$2\alpha = 60°$
$\cos II = 0.5$
$\sin II \approx 0.866$

Area = cos II

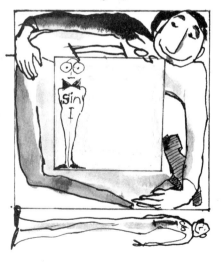

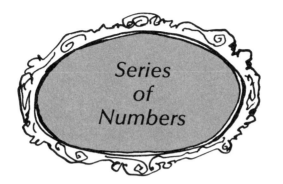

Series
of
Numbers

29

$$1 + 2 + \ldots + n = \frac{n(n + 1)}{2}$$

The tormenting of professors by students is not a new phenomenon, despite what some people would have us believe. In 1787, for example, a teacher overwhelmed by chaos in his classroom decided to punish his students by making them calculate the sum

$$1 + 2 + \ldots + 100$$

of the first 100 whole numbers. Expecting to have calm until the end of the class, the teacher sat down peacefully. Unfortunately for him, one of his students, Carl Friedrich Gauss, announced within five minutes that he had finished the exercise without even doing a single calculation.

"It's all very simple," he explained. "If I write the same sum below the first, but in the opposite order,

$$1 + \ 2 + \ 3 + \ldots + 100$$

$$100 + 99 + 98 + \ldots + 1$$

and then add them together, I obtain twice the number that you asked us to calculate. But $1 + 100 = 2 + 99 = 3 + 98 = \ldots = 101$. By doing this 'vertical' addition, I get 100 times the number 101." As a result, the sum is just

$$\frac{1}{2} \times 100 \times 101 = 5050$$

The professor was already a little lost, but Gauss went on to note that the same approach could be used to calculate the sum of the first n whole numbers:

$$1 + 2 + \ldots + n = \frac{1}{2}[(1 + n) + (2 + n - 1) + \ldots + (n + 1)]$$

Since each of the *n* terms in the expression equals $n + 1$,

$$1 + 2 + \ldots + n = \frac{1}{2} n \, (n + 1)$$

The first student simply adds all the numbers, starting with 1 and working his way up to 100.

Gauss superimposes the two series of numbers and adds them together!

The Fibonacci Sequence:
$F_n = F_{n-1} + F_{n-2}$

Around 1200, the Italian Leonardo Fibonacci considered the following problem. Suppose that there is a fertile pair of rabbits and that they produce a new pair at the end of every month. If each new pair becomes fertile after one month, and if no rabbits die, how many pairs will there be after one year?

To calculate this number, Fibonacci decided to call F_n the number of pairs at the beginning of nth month. Then $F_1 = 1$ and $F_2 = 2$, since at the beginning of the first month there is just the original pair, but at the beginning of the second month the first pair has produced a second pair.

He then noticed that at the beginning of the nth month the pairs can be divided into two groups: a number F_{n-1} of "old" ones, who were already there after $n - 1$ months; and a number of "new" ones, who have just been born. Since a new pair becomes fertile after one month and produces its first descendants after one more month, the number of new pairs is equal to the total number of pairs two months earlier, which is F_{n-2}. As a result,

$$F_n = F_{n-1} + F_{n-2}$$

By using this formula and the initial values $F_1 = 1$ and $F_2 = 2$, it is possible to show that there will be $F_{12} = 233$ pairs after one year. The series of numbers F_n is called the *Fibonacci sequence*. By general agreement, the initial values are normally taken to be 1 and 1 instead of 1 and 2 (so that the following terms in the sequence are shifted):

1, 1, 2, 3, 5, 8, 13, 21, 34, 55, 89, 144, 233, and so on

31
The Number of Ways of Arranging n Objects Is
$$n! = n \times (n-1) \times (n-2) \times \ldots \times 3 \times 2 \times 1$$

If we want to put an algebra book and a geometry book in two different drawers, there are two ways to do it: The algebra book can go in one drawer or the other, and then there is one drawer left for the geometry book. If we add a third drawer and a book of arithmetic, there are now more possibilities. The arithmetic book can go in the first, second, or third drawers, leaving two books to place in the two remaining drawers. We have seen that this can be done in two different ways. As a result, there are $6 = 3 \times 2$ ways to arrange the three books, as shown in the drawing. If we add a fourth drawer and a trigonometry book, there will now be four places to put the new book. For each of these four possibilities, there will be six ways to arrange the remaining three books, as we have just seen. There are therefore

$$24 = 4 \times 6 = 4 \times 3 \times 2$$

ways to arrange the four books.

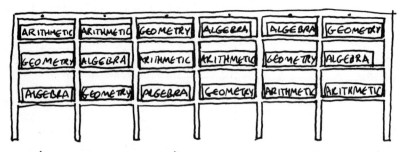

There are three times two ways (6) to arrange the three books.

Of course, 3 × 2 can also be written as 3 × 2 × 1, and 4 × 3 × 2 is equivalent to 4 × 3 × 2 × 1. The number 3 × 2 × 1 is called "3 factorial" and is written as 3! Similarly, 4 × 3 × 2 × 1 is "4 factorial" and is written as 4!

If we now want to know in how many ways the 20 volumes of an encyclopedia can be arranged, we must calculate the number 20! ("20 factorial"):

$$20! = 20 \times 19 \times 18 \times 17 \times 16 \times 15 \times 14 \times 13 \times 12 \times 11$$
$$\times 10 \times 9 \times 8 \times 7 \times 6 \times 5 \times 4 \times 3 \times 2 \times 1$$
$$= 2{,}432{,}902{,}008{,}176{,}640{,}000$$

There are therefore about 2½ quintillion different ways of arranging this encyclopedia in 20 drawers.

"If I do one arrangement per second, it will take me 77 billion years to try all of them!.."

32

$$\frac{1}{2} + \frac{1}{4} + \frac{1}{8} + \ldots = 1$$

Zeno of Elea was a Greek philosopher who loved paradoxes. One day he proclaimed that movement was impossible. He argued that in order for an arrow to reach its target, it must first cover half of the distance, then half of the remaining distance, then half of the distance still remaining, and so on, so that it might appear that the arrow will never actually arrive.

In fact, we know that the arrow will eventually reach its target. This is because the time required to cover successively smaller distances becomes shorter and shorter. Nevertheless, Zeno's paradox shows that a meter can be cut into 1/2 meter (the first half), plus 1/4 meter (the first half of the remaining section), plus 1/8, plus 1/16, and so on. As in the case of the golden ratio in Chapter 17, it is inconvenient to keep calculating the value of this series indefinitely. However, even if we consider only 1/2 (= 0.5), 1/2 + 1/4 (= 0.75), and 1/2 + 1/4 + 1/8 (= 0.875), we can see that the value gets closer to 1 each time that we "add a term." As a result, we can represent this convergence by writing

$$\frac{1}{2} + \frac{1}{4} + \frac{1}{8} + \ldots = 1$$

The three spaced periods indicate that as more and more terms are added to the sum on the left, its value gets closer and closer to 1.

For entertainment, the reader can derive this formula by setting $x = 1/2$ in the equation discussed in Chapter 33.

33

$$1 + x + x^2 + x^3 + \dots = \frac{1}{1-x} \text{ (for } |x| < 1)$$

With the help of Zeno of Elea, we have seen how we can use clever reasoning to deduce the sum of an infinite series of numbers without actually having to do a real calculation. We will now tackle a more general formula of the same type.

Thanks again to our friend who hangs signs, we know that

$$(1 - x)(1 + x) = 1 - x^2$$

If we add the number x^2 to the term $1 + x$ on the left side of this equation, it will be multiplied by 1 to give x^2 and by $-x$ to give $-x^3$, and so

$$(1 - x)(1 + x + x^2) = 1 - x^2 + x^2 - x^3 = 1 - x^3$$

By continuing to add successive terms x^3, x^4, and so on to the term $1 + x + x^2$, the expression on the right will become $1 - x^4$, then $1 - x^5$, and so on.

If x is a number smaller than 1, the numbers x^2, x^3, x^4, and so on in the expression on the right will become smaller and smaller, and they will rapidly approach 0. For example, if $x = 0.2$, then $x^2 = 0.04, x^3 = 0.008, x^4 = 0.0016, x^5 = 0.00032$, and so on. If we continue the process of adding successive terms indefinitely, we will then have

$$(1 - x)(1 + x + x^2 + x^3 + \dots) = 1$$

from which the equation in the title can be derived.*

---

*This equation is equally valid for both positive and negative values of x as long as the absolute value $|x|$ is smaller than 1.

Let's return to the example $x = 0.2$. The value of the term on the right is

$$\frac{1}{1-x} = \frac{1}{0.8} = 1.25$$

whereas the value of the first six terms on the left is

$$1 + x + x^2 + x^3 + x^4 + x^5 = 1.24992$$

If further terms like x^6, x^7, and so on are added, the value of the sum gets even closer to the "true" value 1.25.

34
A Few Other Sums

On the preceding pages, we have seen with the help of Gauss, Zeno of Elea, and our friend who hangs signs that it is relatively easy to calculate particular sums such as the first n whole numbers, $1/2 + 1/4 + 1/8 + \ldots$, and $1 + x + x^2 + \ldots$. Over time, numerous mathematicians have tried to use other "infinite" sums analogous to those in Chapters 21, 32, and 33 in order to calculate certain special numbers, including π, e, and log 2, that we have encountered during the course of our mathematical explorations. Some of the most famous of these formulas appear below. For amusement, the reader can check these equations by calculating the value of the expressions on the right for increasing numbers of terms, and by verifying that the sums get closer and closer to the special numbers on the left. Today, of course, it is much easier to get these special numbers by using a calculator.

$$\frac{\pi}{4} = 1 - \frac{1}{3} + \frac{1}{5} - \frac{1}{7} + \frac{1}{9} - \ldots$$

$$\log 2 = 1 - \frac{1}{2} + \frac{1}{3} - \frac{1}{4} + \frac{1}{5} - \ldots$$

$$\frac{\pi^2}{6} = 1 + \frac{1}{2^2} + \frac{1}{3^2} + \frac{1}{4^2} + \frac{1}{5^2} + \ldots$$

$$e = 1 + \frac{1}{1} + \frac{1}{2} + \frac{1}{6} + \frac{1}{24} + \frac{1}{120} + \ldots$$

In the last sum, the denominators $1, 2, 6, 24, 120, \ldots$ are simply the numbers $n!$ encountered in Chapter 31.

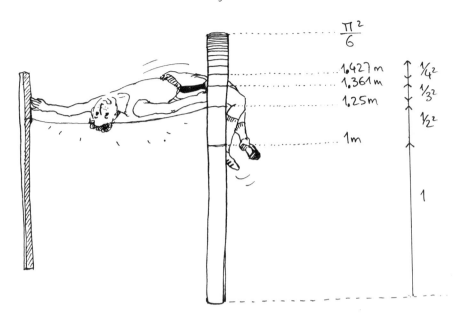

Objects
in
Space

Euler's Theorem:
$f - e + v = 2$

A polyhedron is a solid bounded by flat faces, straight edges, and vertices. For example, the cube and the tetrahedron are polyhedra. All classical polyhedra satisfy Euler's theorem:

$$f - e + v = 2$$

where f is the number of faces, e is the number of edges, and v is the number of vertices. For the cube,

$$6 - 12 + 8 = 2$$

and for the tetrahedron,

$$4 - 6 + 4 = 2$$

Without proving this result in a completely general way, we can nevertheless see that it must be true since a cube can be "transformed" into a tetrahedron by adding or removing edges and vertices. Each of these changes leaves the number $f - e + v$ unchanged. Analogous transformations can interconvert all other classical polyhedra, such as prisms and octahedra.

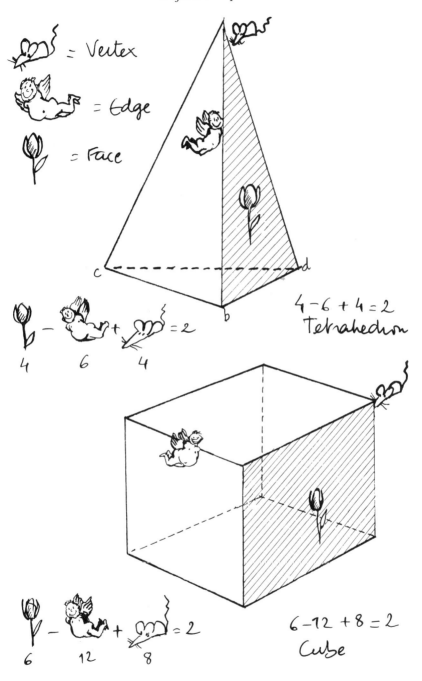

= Vertex

= Edge

= Face

$4 - 6 + 4 = 2$
Tetrahedron

$\quad - \quad + \quad = 2$
4   6   4

$\quad - \quad + \quad = 2$
6   12   8

$6 - 12 + 8 = 2$
Cube

36

The Surface Area of a Sphere
Equals $4\pi R^2$

An impulsive little girl who had just been given a paper globe found nothing better to do with it than to cut it into thousands of tiny pieces. Since she still had the cylindrical container that it came in, she decided to unroll the box and glue the pieces inside. The box was originally big enough for the whole globe, so she assumed that there would be plenty of space.

She was astonished to discover that she needed to use the entire surface to glue all the pieces of the globe. Her impulsiveness did not prevent her from having a flair for geometry, so she realized that the surface area of the cylindrical box must be equal to the surface area of the enclosed sphere. Since the height of the cylinder is twice the radius R of the sphere and since the circumference of its base is $2\pi R$, the area is

$$A = 2R \times 2\pi R$$

$$\boxed{A = 4\pi R^2}$$

Reflecting on her discovery, she concluded that since the radius of the earth is about 6400 kilometers, its surface area must be approximately

$$A \approx 4 \times 3.14159 \times (6400)^2 \ \text{km}^2$$

$$\approx 515 \text{ million square kilometers}$$

By comparison, the United States, which covers about 9.4 million square kilometers, represents less than 2% of the surface of the globe.

37

The Volume of a Sphere Equals
$$\frac{4}{3}\pi R^3$$

We saw in Chapter 10 how Archimedes was able to calculate the area of a circle of known circumference by cutting it cleverly into thin slices. Artifacts recently unearthed by archaeologists now lead us to believe that the ancient Egyptians accomplished something similar by calculating the volume of a sphere of known surface area.* These artifacts were almost certainly made at the time of the construction of the great Pyramids. As the figure shows, the artifacts are uniform small stone pyramids with square bases and sharp summits, and they pack together to form a rough sphere with the summits all located at the center. The volume v of each pyramid is equal to the area a of its base times its height R (equal to the radius of the sphere) divided by 3:

$$v = \frac{1}{3}(R \times a)$$

(Unlike the trapezoid of Chapter 8, the pyramid cannot be readily cut up into geometrically simple pieces whose volume is easy to calculate, so we simply ask the reader to accept this classical formula.)

If we now imitate Archimedes and add up all these little volumes, we get the volume V of the whole sphere, which must therefore be $R/3$ multiplied by the sum of all the areas a, which is just the surface area A of the sphere. Therefore

$$V = \frac{R}{3} \times A = \frac{R}{3} \times 4\pi R^2$$

$$\boxed{V = \frac{4}{3}\pi R^3}$$

---

*As in the case of the circle (Chapter 10), the vocabulary used by mathematicians makes a distinction between a sphere, which is just a surface, and a ball, which is the solid object enclosed by that surface. However, it is common in everyday language to use the word *sphere* to refer to the solid object.

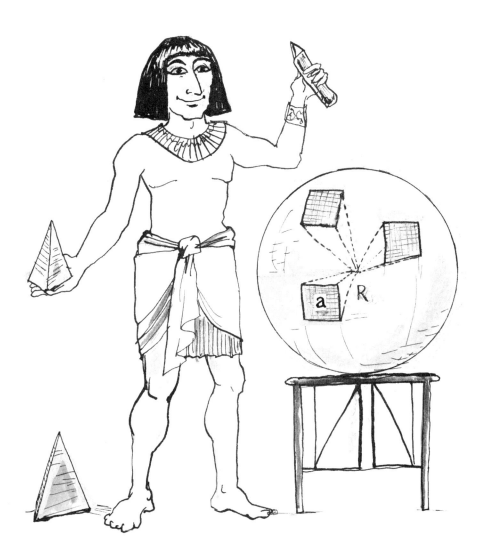

The Angle at the Center of
A Regular Tetrahedron Equals 109° 28′

Some of the stars of the world of mathematics, such as the number π, have been famous since ancient times whereas others, including the numbers i and e encountered in this book, are relatively recent discoveries. It is amusing to find that another famous number, less well-known than π, i, and e, is hidden in the heart of diamonds.

Diamond is made up of a regular arrangement of carbon atoms that lie at the corners of a regular tetrahedron (C_1, C_2, C_3, C_4) and at its center of gravity G.

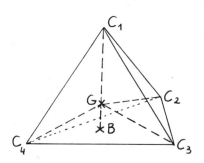

It is possible to show that the angles C_1GC_3, C_1GC_2, and so on are all equal and that their cosine is $-1/3$. This is because if B designates the point where the line C_1G intersects the plane $C_2C_3C_4$, then the distance GB equals one-third of C_1G. If the cosine is $-1/3$, then the angle must be approximately 109° 28′.

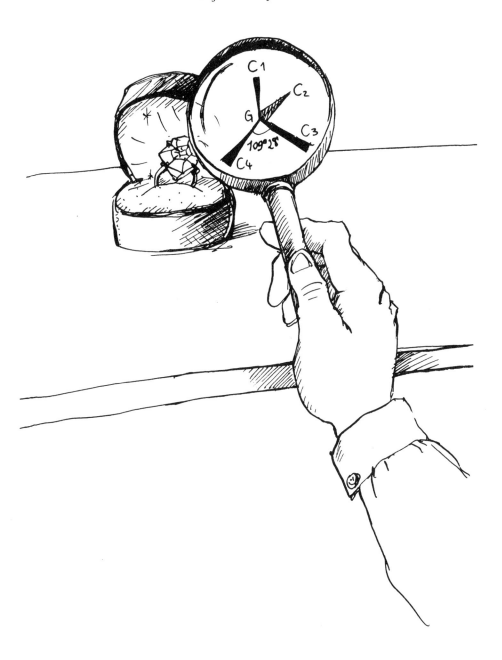

39

The Bridges of Königsberg

In the 18th century, idle citizens of the Prussian city of Königsberg passed the time by strolling over the seven bridges that crossed the Pregel River, perhaps dropping pebbles as they walked like Tom Thumb in the legend and in the drawing on the opposite page. They tried repeatedly, without ever succeeding, to cross all the bridges without passing over any of them more than once. Finally, the great mathematician Euler showed that it could not be done. To do this, he simplified the problem by giving it the graphical representation shown below, in which the points B, C, and D each stand for a bank of the river, A represents the island, and each line corresponds to one of the seven bridges.

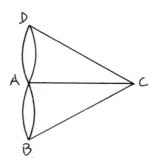

Then Euler noted that the problem of crossing the bridges was equivalent to the problem of finding a circuit that follows all the lines without ever passing over the same one twice. If it were possible, then whenever a point is reached along the circuit, there must be two lines going through it, one corresponding to arriving and the other to leaving. This means that an even number of paths must radiate from each point, except for the two points where the circuit starts and finishes. However, this condition cannot possibly be satisfied, since each point A, B, C, and D is linked to the others by an odd number of lines.

Königsberg later became a city in the U.S.S.R. called Kaliningrad. Today, the problem of the bridges has been eliminated by the construction of an eighth bridge. The reader can confirm that the new bridge, however it is placed, will make it possible to cross all the bridges without ever backtracking.

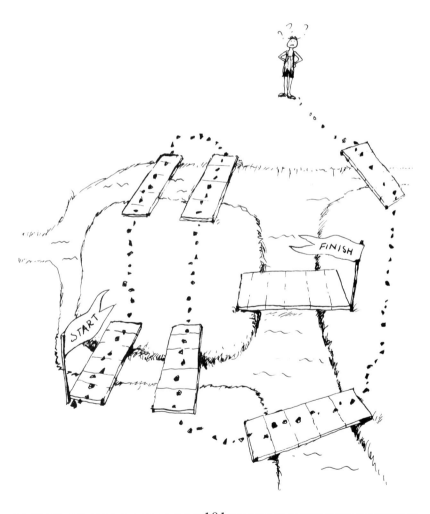

Whole Numbers, Prime Numbers

Lagrange's Theorem:
Every Whole Number Is
the Sum of Four Squares

The "square carpet incident" was at the heart of a recent controversy widely reported in the newspapers. The squabble began when a group of Martian carpet manufacturers decided to sell only square rugs measuring a whole number of meters on each side. That was fine for a customer needing 16 m², for example, since a single piece 4 meters by 4 meters would do, but someone who wanted 20 m² was out of luck. As a result of a new development in this story, the dispute now appears to have been settled in favor of the carpet manufacturers. They uncovered the work of Lagrange, a mathematician who lived during the 18th century and made the following discovery: Any rectangular surface whose area is a whole number can be broken up into 1, 2, 3 or 4 surfaces whose areas are squares of whole numbers. For example:

$$\boxed{\begin{array}{cccc}1&2&3&4\\5&6&7&8\\9&10&11&12\end{array}} = \boxed{\begin{array}{ccc}1&2&3\\5&6&7\\9&10&11\end{array}} + \boxed{4} + \boxed{8} + \boxed{12}$$

$$\boxed{\begin{array}{ccccc}1&2&3&4&5\\6&7&8&9&10\\11&12&13&14&15\end{array}} = \boxed{\begin{array}{ccc}1&2&3\\6&7&8\\11&12&13\end{array}} + \boxed{\begin{array}{cc}4&5\\9&10\end{array}} + \boxed{14} + \boxed{15}$$

In general, Lagrange demonstrated that any whole number can be decomposed into sums of no more than four squares. Another example is:

$$97 = 64 + 25 + 4 + 4 = 8^2 + 5^2 + 2^2 + 2^2$$
$$= 81 + 16 = 9^2 + 4^2$$

as shown in the figure on the opposite page. In this case, there are two ways of breaking up 97 into squares, one more economical than the other.

For further amusement, the reader can pick other numbers and show that they can always be broken up in at least one way to give sums containing no more than four squares.

41

Fermat's Last Theorem

We have seen in Chapter 40 how the square carpet incident stimulated interest in finding ways of breaking numbers up into sums of squares. People quickly noticed that certain squares could be further decomposed into sums of two squares. For example,

$$25 = 16 + 9$$

and

$$169 = 144 + 25$$

Similar observations had already been made in the 17th century and had led a mathematician of that era, Pierre de Fermat, to ask a three-dimensional version of the same question: Can a cube whose sides are a whole number ever be broken up into two other cubes whose sides are whole numbers? In other words, are there any whole numbers x, y, and z (representing the sides of the three cubes) for which

$$x^3 = y^3 + z^3?$$

Unlike squares, no such cubes exist. For example, $8 = 2^3$ or $27 = 3^3$ can never be written as the sum of the cubes of two whole numbers, so the girl in the drawing will never be able to balance her scales.

In an extraordinary achievement, Fermat claimed to have discovered a beautiful proof of an even more general result: The impossibility of breaking cubes up into sums of cubes extends to all whole powers greater than 3, so that if x, y, and z are whole numbers and if $n \geqslant 3$, then the equation

$$x^n = y^n + z^n$$

can never be satisfied. However, even after three centuries of attempts, no other mathematician has been able to establish that this general proposition is true. Any reader who can discover a proof will become famous.

With squares, the scales can be made to balance.

With cubes, the scales can never be made to balance.

42

Prime Numbers Are Indivisible

Editors of the sports sections of newspapers have long been aware that their photographers are always in better spirits when they are sent off to cover rugby matches than when they have to cover soccer. A brief survey has now uncovered the reasons for this preference and has led to some surprising conclusions.

"It has nothing to do with the sports themselves," explains one photographer. "It is simply because in rugby there are 15 players, so in team photos they can be arranged in three rows of five. But in soccer, on the other hand, there are 11 players, and they can't be arranged in even rows!"

A scientific expert has confirmed that this statement makes sense. "In fact," he replies, "the number 15 can be subdivided into smaller whole numbers: $15 = 5 \times 3$. In contrast, the number 11 can only be divided by itself and by 1, so we call it a *prime number*."

He adds that the first ten primes are 2, 3, 5, 7, 11, 13, 17, 19, 23, and 29. With the exception of 2, all of these numbers are odd. However, odd numbers are not necessarily primes. For example, $9 = 3 \times 3$, $15 = 5 \times 3$, $35 = 5 \times 7$, and so on.

Asked how the problem of soccer photographs might be solved, the expert came up with an idea that revealed an inhuman lack of concern for some of the players: "Cut the half-backs in two!" he suggested. But the association of professional soccer players objected, so photographers must continue to take team photos with four players in front and seven behind.

43

Goldbach's Conjecture:
Every Even Number
Is the Sum of Two Primes

In this book, we have encountered numerous mathematical results and formulas, and we have seen that the motivations leading to their discovery have been extremely diverse. In number theory, we enter a field where pure curiosity and the simple desire to know more are the principal motivations. This is what led Goldbach, a German mathematician living in Russia around 1750, to notice that

$$4 = 2 + 2, 6 = 3 + 3, 8 = 5 + 3, 10 = 7 + 3, 12 = 7 + 5,$$

$$14 = 7 + 7 = 3 + 11, 16 = 11 + 5 = 3 + 13,$$

$$18 = 11 + 7 = 13 + 5, 20 = 13 + 7 = 17 + 3$$

In other words, every even number from 4 to 20 is the sum of two primes. Goldbach then asked himself the following question: Is this true for all even numbers?

The reader can confirm that for any even number N he picks, there are always two prime numbers a and b such that $N = a + b$. For example, $48 = 11 + 37$. Nevertheless, no mathematician has ever succeeded in proving Goldbach's conjecture in all its generality. No one has ever found an even number that cannot be expressed as the sum of two primes, but no one has shown that this might not happen someday.

44

The Prime Number Theorem

If you look at a table of prime numbers, you will see that they are distributed very irregularly. For example, there can be a long series of numbers that are not prime, such as 90, 91, 92, 93, 94, 95, and 96, followed by several prime numbers very close to one another, such as 97, 101, and 103. More evidence that the distribution is irregular comes from the fact that no one has ever found a simple "formula" guaranteed to generate prime numbers.*

Near the end of the last century, however, mathematicians arrived at the following remarkable conclusion: Although it may be impossible to find a general formula that gives the exact value of the nth prime number, its order of magnitude can nevertheless be estimated. As n becomes larger and larger, the nth prime number gets closer and closer to $n \log n$.

For example, if $n = 100$, the 100th prime number is 541, whereas the value of $100 \log 100$ is approximately 460.517. There is therefore a difference of about 81 between the true value and the approximation, which corresponds to a relative error of $81/541 \approx 15\%$. If $n = 1000$, however, the 1000th prime number is 7919 and the approximate value is 6907.75. In this case, the relative error is only $1011/7919 \approx 12.8\%$. Proof of the so-called *prime number theorem* is extremely difficult, but these examples suggest that the relative error does in fact become closer and closer to 0 as n becomes larger and larger:

$$\boxed{n\text{th prime number} \approx n \log n \ (n \text{ large})}$$

---

*"Formulas" that give all of the prime numbers do exist, but they involve such complex calculations that they are of no practical value.

Chance

The Chance of Winning
the Lottery

A "6/49" loto, which is a lottery popular in Québec and in Europe, involves drawing 6 numbers at random out of 49. The big winners are the ones who have picked the right 6 numbers on their lottery tickets. There are obviously many ways to choose 6 numbers out of 49, which is why the chances of winning big are very small. For a player who has picked 6 numbers, what exactly are the odds on winning?

To have an idea of the problem involved, let's imagine a simplified "2/5" lottery where 2 numbers are drawn out of 5. The following pairs can be drawn,

$$(1,2) \ (1,3) \ (1,4) \ (1,5) \ (2,3) \ (2,4) \ (2,5) \ (3,4) \ (3,5) \ (4,5)$$

giving a total of 10. This is because there are 5 possibilities for the first number, then 4 remaining possibilities for the second, giving a total of 20 possible draws. But each draw is actually counted twice; for example, (1,2) and (2,1) are really the same since the order of the numbers doesn't matter. As a result, there are in fact $20/2 = 10$ possible outcomes. By a similar argument, it is possible to show that the number of different draws in a real 6/49 lottery is

$$N = \frac{49 \times 48 \times 47 \times 46 \times 45 \times 44}{6 \times 5 \times 4 \times 3 \times 2 \times 1}$$

In this case, the numerator takes the place of 5×4 and the denominator replaces 2×1. The number of draws is therefore

$$N = 13,983,816$$

Unlike the player of the 2/5 lottery watching the results televised on the opposite page, players of a real 6/49 lottery cannot afford to buy enough tickets to cover most of the possible combinations. In fact, the probability p of winning is only 1/13,983,816 or

$$p \approx 0.000000072$$

which corresponds to 7 chances in 100 million! That's why so few people win big.

46

Roulette and d'Alembert's Martingale

If you are not too ambitious, there is a sure way to win at roulette: the strategy called d'Alembert's Martingale. This is a betting system that promises winnings of $1 for an initial bet of $1.

You begin by betting $1 on red. If red comes up, you win $2 for an initial bet of $1; your net winnings are therefore $1 and you can quit. If not, you then bet $2 on red. If red comes up, you win $4 after having bet a total of $3 ($1 in the first round and $2 in the second); your overall winnings are therefore $1 and you can quit. If not, you bet $4 on red. If red comes up, you win $8 after having bet a total of $7 ($1 + 2 + 4$); you therefore win $1. If not, you continue to follow the same strategy, each time doubling your bet on red, until red finally comes up. Since this must eventually happen, the betting strategy is guaranteed to work.

The problem is that in order to win $1000, you may need to wager sums of $3000, $7000, or even $15,000 or more. Besides the fact that you may not necessarily have this much money,* bets in roulette are limited. This means that you may not be able to follow the strategy at the very moment that red decides to come up, which is what has happened to the unhappy bettor on the facing page, who has lost the sum of all his bets. On the other hand, if you only want to win enough to be able to say you beat the house, the strategy is perfectly suitable.

---

*Note that a bettor with an initial amount of $2^n - 1$ dollars at his disposal will be bankrupt if black comes up n times in a row!

Pascal's Triangle

We have seen in Chapter 45 that there are 10 ways to choose 2 objects from a group of 5. It is also possible to analyze the numbers of ways to choose 1, 3, 4, or 5 objects from the same group. In these cases, there are 5, 10, 5, and 1 different combinations. The reader will note that choosing 2 objects is equivalent to choosing the 3 objects that remain, so it is reasonable that there should also be 10 ways to choose 3 objects from a group of 5.

The numbers of different possibilities can also be found by using a "geometric" device called *Pascal's triangle*. At the top we place the number 1 all by itself. On the second line, we place two new 1's on either side of the first 1. On succeeding lines, each number that appears is calculated by adding together the two numbers that appear on either side of it on the line above. The two ends of each line of numbers are always 1's. The third line is therefore 1, 2 (= 1 + 1), 1, and the fourth is 1, 3 (= 1 + 2), 3 (= 2 + 1), 1. Continuing in this way, we find that the sixth line is 1, 5, 10, 10, 5, 1, as already deduced above.

One of the things that makes this triangle remarkable is that the numbers on the nth line provide the coefficients in the formulas for expanding $(a+b)^{n-1}$:

$$(a+b)^2 = 1a^2 + 2ab + 1b^2$$
$$= a^2 + 2ab + b^2 \text{ (see Chapter 6)}$$
$$(a+b)^3 = 1a^3 + 3a^2b + 3ab^2 + 1b^3$$
$$= a^3 + 3a^2b + 3ab^2 + b^3$$
$$(a+b)^4 = 1a^4 + 4a^3b + 6a^2b^2 + 4ab^3 + 1b^4$$
$$= a^4 + 4a^3b + 6a^2b^2 + 4ab^3 + b^4$$
$$(a+b)^5 = 1a^5 + 5a^4b + 10a^3b^2 + 10a^2b^3 + 5ab^4 + 1b^5$$
$$= a^5 + 5a^4b + 10a^3b^2 + 10a^2b^3 + 5ab^4 + b^5$$

Today
and
Tomorrow...

The Binary System: 1 + 1 = 10

Our habit of using the so-called "decimal" system for numerical calculations starts in childhood, and we consider it perfectly natural to count in this way. However, it requires a significant amount of memorization (for example, to learn the multiplication tables), and in fact it is not as practical as it seems to be. For this reason, computers are programmed to use a different method of calculation called the "binary" system, in which the number 2 takes the place of the number 10. In this system, all numbers are written using only the symbols 0 and 1. The following table gives the binary and decimal representations of the first 16 whole numbers and shows how successive numbers are created in the binary system:

Decimal	Binary	Decimal	Binary
0	0	8	1000
1	1	9	1001
2	10	10	1010
3	11	11	1011
4	100	12	1100
5	101	13	1101
6	110	14	1110
7	111	15	1111

In this new system, addition is very simple and is governed by the following three rules:

$$
\begin{array}{cc}
0 & 0 \\
+0 & +1 \\
\hline
0 & 1
\end{array}
\quad \text{or} \quad
\begin{array}{cc}
1 & {}^{1}1 \\
+0 & +1 \\
\hline
1 & 10
\end{array}
$$

The small ^{1} in the last addition indicates that 1 is carried over from the first column.

According to the last rule, the conventional equality $1 + 1 = 2$ now can be written as $1 + 1 = 10$, as shown in the title. Another example is

$$
\begin{array}{r}
101 \\
+\,110 \\
\hline
1011
\end{array}
$$

which corresponds to the conventional addition $5 + 6 = 11$.

It is amusing to note that the binary system, which has recently come into prominence with the introduction of modern methods of computation, is in fact an old discovery that probably dates back to the German mathematician Leibniz. In a letter dated February 26, 1701, now in the collection of the French Academy of Sciences, Leibniz wrote the following lines:

> I enclose an attempt to devise a numerical system that may prove to be completely new. Briefly, here is what it is… By using a binary system based on the number 2 instead of the decimal system based on the number 10, I am able to write all numbers in terms of 0 and 1. I have done this not for mere practical reasons, but rather to allow new discoveries to be made… This system can lead to new information that would be difficult to obtain in any other way….

When he wrote these lines, Leibniz could hardly have imagined that among the "new discoveries" would be the computer and the word-processing techniques that we use today to cite his work.

49
Toward Infinity

During this exploration of the world of mathematics, we have frequently made statements like "this sum gets closer and closer to... as more and more terms are calculated" or "the number... becomes smaller and smaller as n becomes larger and larger." As an example of the first statement, we have written that

$$\frac{1}{2} + \frac{1}{4} + \frac{1}{8} + ... = 1$$

Behind all of these statements lies the barely hidden idea of infinity. The preceding equation is equivalent to saying that if the sum of an infinite series of numbers 1/2, 1/4, 1/8, and so on is calculated, the result is 1. In practical terms, of course, the calculation is impossible since an infinite sum goes beyond the finite reach of man. Moreover, even if we actually had all of eternity, we would still never be able to "reach" infinity.

Over the course of centuries, mathematicians have nevertheless learned to deal with the difficult notion of infinity. Being unable to reach infinity themselves, they have tried to bring infinity a little closer to home. They have therefore made it an integral part of the world of mathematical concepts, a magical land where objects prove to be useful even if all of their properties are not fully understood. As a result, infinity is no longer a paradoxical concept that makes us feel uneasy. On the contrary, even if infinity is still an idea that must be handled with care, it is a tremendous source of rich discoveries.

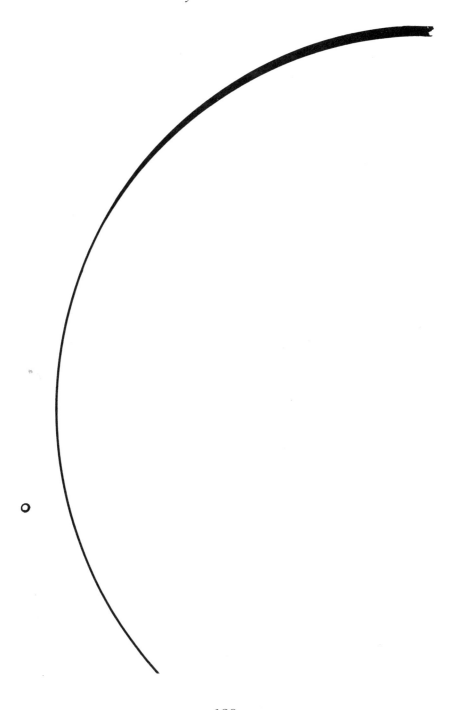

Annex

In stories involving real characters, we have occasionally been led to alter the historical truth. The goal of the additional information provided below is to set the record straight:

— There are many different proofs of the Pythagorean theorem (Chapter 8). The one we have chosen, which was in fact devised by President James Garfield, stands out because of its originality and simplicity. On page 137, we provide a more classical proof of the theorem that also makes use of a geometric construction.

— Archimedes actually discovered the method described in Chapter 10 for calculating the area of a circle. On the other hand, the large cake existed only in the imagination of the authors.

— Even though Greek mathematicians did not have the formulas that appear at the end of Chapter 16, they were nevertheless able to calculate the roots of quadratic equations.

— John Napier, the Scottish nobleman of the 17th century, actually employed a gardener. Otherwise, however, the stories presented in Chapters 19 to 24 are totally fictitious. The major contribution of Napier to the study of logarithms was the creation of a table of numerical values in 1614. Other results presented in this section actually came from the work of later mathematicians, including Newton and Euler.

— The conversation involving Euler, Diderot, and Catherine the Great of Russia really took place. The footnote in Chapter 27 specifies what was actually said on this occasion.

— It is difficult to guarantee that Gauss discovered the simple formula for calculating the sum $1 + 2 + ... + 99 + 100$ while he was merely a schoolboy, but this is what we are told. At the very least, the story can be considered to be an "authentic legend."

— Chapter 30 shows how to calculate the number of pairs of rabbits produced at the end of one year by a single fertile pair, if each fertile pair produces one new pair at the end of one month and the new pair becomes fertile at the end of one month. Leonardo Fibonacci actually used this calculation in 1202 in *Liber abaci* to introduce the sequence that now bears his name.

Finally, we have made up the stories about farmers who figure out the area of rectangles or triangles (Chapters 3 and 4) or about children who discover well-known formulas by cutting things up in clever ways (Chapters 6 and 7), but it is likely that similar or even identical events have actually taken place. *Trifolium giganteum* (Chapters 3, 4, and 5), and dwarf contortionists (Chapters 11 and 28) do not actually exist, while the stories about the race for the Golden Gouda (Chapter 14) and the Egyptian artifacts (Chapter 37) are little tales that are historically or technologically improbable. Obviously, the square carpet incident (Chapter 40) could only have happened on the planet Mars, and the sour mood of the soccer photographers (Chapter 42) would only have lasted as long as the blink of a shutter.

Biographical Index

d'Alembert (Jean Le Rond) (1713–1783): French philosopher and mathematician who was the scientific adviser for the *Encyclopedia*. He is known in mathematics for his work in analysis and mechanics. His name, along with that of Gauss, is associated with the fundamental theorem of algebra, which states that every polynomial equation has at least one root, which is a complex number.

Archimedes (287–212 B.C.): Greek mathematician, physicist, and engineer who lived in Syracuse in the 3rd century B.C. Creator of the method of calculation presented in Chapter 10, he was one of the great thinkers of antiquity. He also discovered the principle of buoyancy, which explains the flotation of boats and other phenomena. He was killed when Syracuse fell to the Romans after a siege that had been prolonged for several years thanks to the machines he had devised as a military engineer.

Cardano (Girolamo) (1501–1576): Italian algebraist and doctor who discovered formulas for solving cubic equations. His work marked the first appearance of the square roots of negative numbers, which would later become "imaginary" numbers (Chapter 18). He is also given credit for designing prototypes of the system of joints that now bears his name and is used to control the steering and transmission of automobiles. In addition to his work in medicine, he practiced astrology and produced horoscopes, for which he was condemned by the Inquisition for heresy.

Catherine II of Russia, named Catherine the Great (1729–1796): Empress of Russia from 1762 to 1796, Catherine the Great helped bring about the unification of her country. Instilled with liberal fervor, she attempted at the beginning of her reign to change society by granting greater political freedom. A number of philosophers and scientists, including Euler and Diderot (Chapter 27) as well as Voltaire, benefited from her patronage.

Christophe (pseudonym of *Georges Colomb*) (1856–1945): French naturalist who was assistant director of the laboratory of botany at the Sorbonne. His written work consists of various handbooks of natural science as well as a trilogy of illustrated texts entitled *The Fenouillard Family, Camembert the Fireman,* and *The Obsession of Cosine the Scholar.*

Diderot (Denis) (1713–1784): Great French philosopher and writer. The minor role that he plays in our book fails to do justice to his important contributions. He appears not to have been extremely gifted in mathematics, and it is true that Euler took advantage of this weakness to make fun of him in front of Catherine the Great of Russia. His atheism, which prompted Euler's quip, caused him to be imprisoned in 1749. In addition to his novels (*Jacques the Fatalist, Rameau's Nephew,* and *The Nun*), he and d'Alembert were prime forces behind the *Encyclopedia.*

Euler (Leonhard) (1707–1783): Swiss mathematician and physicist. By virtue of his extraordinary powers of creativity, he discovered and developed entirely new fields of mathematics and went more deeply into other subjects already studied by his predecessors. In this book, we have encountered him several times in connection with Euler's constant (Chapter 21), imaginary exponents (Chapter 27), polyhedra (Chapter 35), and the problem of the bridges of Königsberg (Chapter 39). The variety of these encounters is a good indication of the remarkable breadth of his work. A complete edition of his work was published at the beginning of the 20th century and consists of almost 100 volumes.

Fermat (Pierre de) (1601–1665): French lawyer from Toulouse who contributed to the modernization of calculus. His greatest fame comes from his work on the theory of numbers, a field that had remained dormant in the West for more than a thousand years. It was common in Fermat's time to announce results without actually publishing their proof. In effect, this challenged other mathematicians to show that they were clever enough to derive the proof themselves, and Fermat played this game frequently. One of his results, the "last theorem" presented in Chapter 41, has continued for more than three hundred years to withstand all efforts of other mathematicians to discover a proof.

Fibonacci (Leonardo), named Leonardo of Pisa (born around 1175, died after 1240): Italian mathematician who invented the sequence that now bears his name (Chapter 30). His major contribution was the publication in 1202 of *Liber abaci,* which brought the remarkable mathematical achievements of the Arabs to the attention of the western world. In particular, this book introduces zero and the numbers, now called Arabic numerals, that we use every day.

Garfield (James) (1831–1881): American politician and president. Self-taught, he was an amateur mathematician and a professor of ancient languages. After having fought for the Union in the Civil War, he became leader of the Republican Party in 1876. He was elected President of the United States in November 1880 and took office in January 1881 as prescribed by the U.S. constitution. Two months later, he was assassinated.

Gauss (Carl Friedrich) (1777–1855): German mathematician and physicist. Nicknamed "princeps mathematicorum" (prince of mathematicians) by his contemporaries, Gauss reigned supreme over all branches of mathematics for thirty years. There is no field that his universal genius failed to influence, and his record as an innovator is unmatched. He accomplished so much that even today textbooks are filled with "Theorems of Gauss" covering an extremely wide range of fields.

Goldbach (Christian) (1690–1764): German mathematician. He is known primarily for the conjecture presented in Chapter 43, which states that every even number is the sum of two prime numbers. In a letter written in 1742 while he was living in Russia, he brought his conjecture to the attention of Euler. Even today, the conjecture still remains unproved.

Lagrange (Joseph Louis) (1736–1813): French mathematician born in Turin. He worked on a broad range of problems in mathematics, particularly in analysis and number theory, and discovered the "theorem of four squares" discussed in Chapter 40. He was involved in a variety of activities and held posts as professor of mathematics at the École Polytechnique and as president of the Committee on Weights and Measures, which was established in 1795.

Leibniz (Gottfried Wilhelm) (1646–1716): German philosopher and mathematician. At the same time as Newton, Leibniz independently discovered the principles of differential and integral calculus (Chapter 25). In the course of this work, he devised mathematical notations still used today, including the symbol $\int$ that designates an integral. We have seen elsewhere (Chapter 48) that Leibniz was the first to think of using the binary system in mathematics. In his philosophical essays, he used the notion of "monads" to develop an optimistic view of the world that provoked numerous responses later in the 18th century, particularly from Voltaire and Diderot.

Napier (John), Laird of Merchiston (1550–1617): Scottish mathematician. He gave his name to "natural" logarithms and published a table of their numerical values in 1614. In addition to his mathematical activities, he appears to have been obsessed with black magic, which gave him an aura of mystery and made him feared by his servants and neighbors.

Newton (Isaac) (1642–1727): English scholar. At the same time as Leibniz, Newton independently invented differential and integral calculus, and he was the originator of modern analysis. His most famous discovery, the law of universal gravitation, was inspired according to legend by apples falling in his garden. In addition, "Newtonian mechanics" govern the motion of objects moving at much less than the speed of light (as opposed to "relativistic mechanics" created by Albert Einstein at the beginning of the 20th century).

Pascal (Blaise) (1623–1662): French writer and mathematician. He is primarily famous for having invented the first computing machine when he was only eighteen years old. Pascal also helped create the science of probability, and he is recognized for various contributions to the study of conic sections. Converted in 1654 to Jansenism, a harsh religious doctrine founded primarily on predestination, Pascal spent the second half of his life defending Jansenism against the attacks of the Jesuits. His principal work, the famous *Pensées*, appeared after his death. He intended it to be a "Defense of Christianity," and it contains the celebrated "Pascal's bet" that since there is a finite possibility of winning eternal happiness by leading a religious life, "It is better to believe in God."

Pythagoras (6th century B.C.): Greek mathematician. He is universally known for the theorem that bears his name (Chapter 8), but he is not actually the discoverer; in fact, Babylonian geometricians were familiar with the principle a thousand years earlier. However, Pythagoras appears to have been the first to establish a number of other properties of triangles, including the rule that the sum of the angles always equals 180° (Chapter 5).

The following argument provides another geometric proof of the Pythagorean theorem that is more classical than the one presented in Chapter 8. The reader can easily confirm that the area of the large square is equal to $(a+b)^2$, since the sides are $a+b$. At the same time, however, the area is also equal to that of a square of side c, plus four right triangles, giving a total of $c^2 + 2\,ab$. If $(a+b)^2$ is expanded according to the formula of Chapter 6, it is possible to establish that $a^2 + b^2 = c^2$.

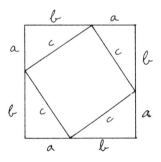

Zeno of Elea (5th century B.C.): Greek philosopher born in Elea. He is known by mathematicians for his paradoxes that appear to demonstrate that movement is impossible. For example, Chapter 32 discusses the paradox involving an arrow heading toward its target. In a similar way, Zeno "proved" that Achilles would never be able to catch up with a tortoise. These paradoxes are among the first examples of efforts to think about infinity, and they have helped prompt mathematicians to develop a deeper understanding of this elusive concept.

Index

About the Authors

Lionel Salem, *Research Professor in the French National Research Center (C.N.R.S.)*, is an internationally renowned theoretical chemist with a special flair for presenting science to the public. In 1979, he published *Marvels of the Molecule* (InterEditions), which received the Glaxo Prize for popular science and has been widely translated. The year 1990 marked the publication of the *Dictionary of Sciences* (Hachette), which was compiled under his editorial direction with the collaboration of thirty leading scientists.

Frédéric Testard, a graduate of one of the Écoles Normales Supérieures, now teaches mathematics at the Université de Nice. With Rached Mneimne, he is coauthor of the book *Introduction to the Theory of Classical Lie Groups* (Hermann), and he collaborated in the compilation of the *Dictionary of Sciences*.

Coralie Salem has a degree in graphic design, and specializes in illustrations and corporate logos. She is the daughter of Lionel Salem and the granddaughter of Raphaël Salem, the noted mathematician.